D1413526

Prentice Hall Series in Process Pollution and Control Equipment

Water Treatment and Waste Recovery
Advanced Technology and Applications

Nicholas P. Cheremisinoff
Paul N. Cheremisinoff

P T R Prentice Hall
Englewood Cliffs, New Jersey 07632

Library of Congress Cataloging-in-Publication Data

Cheremisinoff, Nicholas P.
     Water treatment and waste recovery : advanced technology and
applications / Nicholas P. Cheremisinoff, Paul N. Cheremisinoff.
       p.   cm.
     Includes index.
     1. Water—Purification   2. Sewage—Purification.
I. Cheremisinoff, Paul N.   II. Title.
TD430.C459   1993
628.1'62—dc20   92-39766
                                                                      CIP

Prentice Hall Series in Process Pollution and Control Equipment:

Carbon Adsorption for Pollution Control
Compressors and Fans
Pumps and Pumping Operations
Filtration Equipment for Wastewater Treatment
Heat Transfer Equipment
Lead: A Guidebook to Hazard Detection, Remediation, and Control
Materials and Components for Pollution and Process Equipment

Acquisition Editor: Mike Hays
Editorial/Production Supervision: WordCrafters Editorial Services, Inc.
Cover Designer: Wanda Lubelska
Prepress Buyer: Mary McCartney
Manufacturing Buyer: Margaret Rizzi

 © 1993 by P T R Prentice-Hall, Inc.
A Simon & Schuster Company
Englewood Cliffs, New Jersey 07632

The publisher offers discounts on this book when ordered in bulk quantities. For
more information, write: Special Sales/Professional Marketing, Prentice Hall,
Professional & Technical Reference Division, Englewood Cliffs, NJ 07632.

Printed in the United States of America
10   9   8   7   6   5   4   3   2   1

ISBN 0-13-285784-7

Prentice-Hall International (UK) Limited, *London*
Prentice-Hall of Australia Pty. Limited, *Sydney*
Prentice-Hall Canada Inc., *Toronto*
Prentice-Hall Hispanoamericana, S.A., *Mexico*
Prentice-Hall of India Private Limited, *New Delhi*
Prentice-Hall of Japan, Inc., *Tokyo*
Simon & Schuster Asia Pte. Ltd., *Singapore*
Editora Prentice-Hall do Brasil, Ltda., *Rio de Janeiro*

Contents ━━━━━

Preface

This book summarizes advanced water, wastewater, and separation technologies. The emphasis is on developmental as well as commercial processes and topics in separation and disinfection-sterilization areas. Not all of the technologies are fully developed, and more work is needed to develop new and better science and applications.

Interest in the technologies presented may be applied in pollution control. Barriers to use and commercialization are most frequently expressed by industrial personnel due to the lack of familiarity by management and technical staff. This makes it difficult to apply such processes in cases where they may offer advantages over more conventional methods.

Successful, high-yield, and high-efficiency energy requirements, materials utilization, reactant and product recovery, pollution, and environmental effects are among the factors that contribute to process acceptability. Energy requirements and separation costs can also be critical barriers. The material presented is at a technical level for the manager as well as the practitioner that enables them to evaluate each technology for applicability and potential use.

The areas defined in the materials presented here are those which may be most compatible with developing new technologies in pollution control. Identified also in this study are possible future applications which may encourage their consideration, use, and investigation.

Nicholas P. Cheremisinoff
Paul N. Cheremisinoff

1

Physical Separation Processes

Through the processes of advanced electrodialysis (ED), reverse osmosis (RO), ultrafiltration (UF), and ultracentrifugation (UC), dissolved substances and/or finely dispersed particles can be separated from liquids. Electrodialysis, reverse osmosis, and ultrafiltration rely on membrane transport, the passage of solutes or solvents through thin, porous polymeric membranes. Ultracentrifugation depends on high centrifugal forces to separate dual liquids. A comparison of the characteristics of the four classes of separation processes is given in Table 1.1.

SEPARATION PROCESSES

Electrodialysis

The principle of electrodialysis is that electrical potential gradients will make charged molecules diffuse in a given medium at rates far greater than attainable by chemical potentials between two liquids as in conventional dialysis. When a DC electric current is transmitted through a saline solution, the cations migrate toward the negative terminal, or cathode, and the anions toward the positive terminal, the anode. By adjusting the potential between the terminals or plates, the electric current and, therefore, the flow of ions transported between the plates can be varied.

TABLE 1.1   CLASSES OF SEPARATION PROCESSES

Name	Feed	Separating Agent	Products	Separation Principle	Application Example
Electrodialysis	Liquid	Anionic and cationic membranes; electric field	Liquids	Tendency of anionic membranes to pass only anions, and so on	Desalination of brackish waters
Reverse Osmosis	Liquid solution	Pressure gradient (pumping power) + membrane	Two liquid solutions	Different combined solubilities and diffusivities of species in membrane	Sea water desalination
Ultrafiltration	Liquid solution containing large molecules or colloids	Pressure gradient (pumping power) + membrane	Two liquid phases	Different permeabilities through membrane (molecular size)	Wastewater treatment; protein concentration; artificial kidney
Ultracentrifugation	Liquid	Centrifugal force	Two liquids	Pressure diffusion	Separation of large polymeric molecules according to molecular weight

Electrodialysis can be applied to the continuous-flow type of operation needed in industry. Multimembrane stacks can be built by alternately spacing anionic- and cationic-selective membranes.

Among the technical problems associated with the electrodialysis process, concentration-polarization is perhaps the most serious. Other problems in practical applications include membrane scaling by inorganics in feed solutions as well as membrane fouling by organics. Efficient separation or pretreatment in the influent streams can include activated carbon absorption to reduce or prevent such problems.

Principal applications of electrodialysis include:

- Recovery of materials from liquid effluents, such as processes related to conservation, cleanup, concentration, and separation of desirable fractions from undesirable ones
- Purification of water sources
- Effluent water renovation for reuse or to meet point source disposal standards required to maintain suitable water quality in the receptor streams

Treatment of brackish waters in the production of potable supplies has been the largest application of electrodialysis.

Costs associated with electrodialysis processes depend on such factors as the total dissolved solids (TDS) in the feed, the level of removal of TDS (percent rejection), and the size of the plant. In brackish water treatment, operating costs for very large ED installations (on the order of millions of gallons a day) have been between 40 cents to 50 cents per 1,000 gallons for brackish feed waters, which compares favorably with RO costs.

A rule of thumb for the energy requirements for demineralizing 1,000 gallons of salt water by ED in large capacity plants (4 mgd) is 5 to 7 kWh per 1,000 ppm of dissolved solids removed. Since the efficiency of electrodialytic demineralization decreases rapidly with increasing feed concentrations, this process is best utilized for treatment of weakly saline (brackish) waters containing less than 5,000 ppm of total dissolved solids. In fact, for waters at the low-concentration end of the brackish scale, ED may be the most cost-effective process of all.

Electrodialysis is widely used in the United States in the dairy industry, namely in the desalting of cheese whey. Electrical requirements may vary from 5 to 14 kWh per pound of product solids. Another application of ED is the sweetening of prepared citrus juices. Other less extensive uses of electrodialysis in commercial operations in the United States include tertiary or advanced treatment of municipal sewage water and treatment of industrial wastewaters such as metal-plating baths, metal-finishing rinse waters, wood-pulp wash water, and glass-etching solutions. Potential applications of ED are many. A particular advantage of the electrodialysis process is its ability to produce solutions of high concentrations of soluble salts. A combination of electrodialysis with conventional evaporation, for example, may be substantially cheaper than evaporation alone for the production of dry salt from saline solutions. Competing technologies include reverse osmosis and crystallization.

Reverse Osmosis

Reverse osmosis is fundamentally a means for separating dissolved solids from water molecules in aqueous solutions through membranes composed of special polymers which allow water molecules to pass through while holding back most other types of molecules. In the RO process the feed stream is split into a purified portion (the product water or permeate) and a smaller portion called the concentrate, containing most of the impurities in the feed stream. The percentage of product water obtained from the feed stream is termed the *recovery*. Some important advantages of the reverse osmosis process include:

- Low energy consumption. Because no change of phase is involved (as in distillation), the principal source of energy consumed is electrical to drive pumps.
- Relatively simple processing equipment, resulting in low-to-moderate equipment costs.
- Operations at ambient temperature minimize scale and corrosion problems.

Major problems inherent in RO systems applications include the presence of particulate and colloidal matter in feed water, precipitation of soluble salts, and problems associated with the physical and chemical composition of the feed water.

Applications of large-scale reverse osmosis systems include:

- Upgrading (desalination) of brackish feed waters to produce quality water supplied for municipal and/or industrial requirements or resort communities.
- High-purity rinse water for the solid-state electronic components manufacturing industry.
- Production of boiler feed and process water.

Reverse osmosis appears to have the economic lead in relation to energy consumption for the treatment of all but the most saline waters. Energy costs for RO vary between 4 cents to 32 cents per 1,000 gallons of product. Typical energy consumptions for treatment of brackish wastes fall in the approximate range of 7 to 12 kWh per 1,000 gallons. Comparable energy costs for the most efficient desalination plants employing distillation range between 25 cents to 40 cents per 1,000 gallons. The costs of RO increase with increasing feed concentration and operating pressure and decrease with increasing efficiency of recovery (rejection).

Other smaller-capacity reverse osmosis applications include:

- Treatment of industrial wastes, including metal-finishing effluents, plating rinses, and refinery wastewater.
- Treatment of wastewater and the retrieval of protein substances from fish and shellfish processing plants.
- Processing of pulp and paper mill effluents for reuse.
- Waste treatments in food processing in the dairy products industry in which edible protein and lactose are recovered from cheese whey.
- Desalination (including boron removal) of irrigation return flows.
- Food concentration.

Ultrafiltration

In ultrafiltration, a feed emulsion is introduced into and pumped through a membrane unit; water and some dissolved low molecular weight materials pass through the membrane under an applied hydrostatic pressure. In contrast to ordinary filtration, there is no build-up of retained materials on the membrane filter.

A variety of synthetic polymers, including polycarbonate resins, substituted olefins, and polyelectrolyte complexes, has been employed for ultrafiltration membranes. Many of these membranes can be handled dry, have superior organic solvent resistance, and are less sensitive to temperature and pH than cellulose acetate, widely used in RO systems.

In UF, molecular weight (MW) cutoff is used as a measure of rejection. However, shape, size, and flexibility are also important parameters. For a given molecular weight, more rigid molecules are better rejected than flexible ones. Ionic strength and pH often help determine the shape and rigidness of large molecules. Operating temperatures for membranes can be correlated generally with molecular weight cutoff. For example, maximum operating temperatures for membranes with 5,000–10,000 MW cutoffs are about 65°C. For a 50,000–80,000 MW cutoff, maximum operating temperatures are in the range of 50°C.

The largest industrial use of ultrafiltration is the recovery of paint from water-soluble coat bases (primers) applied by the wet electrodeposition process (electrocoating) in auto and appliance factories. Many installations of this type are operating around the world. The recovery of proteins in cheese whey (a waste from cheese processing) for dairy applications is the second largest application, where a market for protein can be found (for example, feeding cattle and farm animals). Energy consumption at an installation processing 500,000 pounds per day of whey would be 0.1 kWh per pound of product.

Another large-scale application is the concentration of waste-oil emulsions from machine shops, which are produced in association with cooling, lubrication, machining, rolling heavy metal operations, and so on. Ultrafiltration is a preferred alternative to the conventional systems of chemical flocculation and coagulation followed by dissolved air flotation. Ultrafiltration provides lower capital equipment, installation, and operating costs. Ultrafiltration of corrosive fluids such as concentrated acids and ester solution is an important application. The chemical inertness and stability of ultrafilters make them particularly useful in the cleaning of these corrosive solutions. Uses include separation of colloids and emulsions, and recovery of textile sizing chemicals. Biologically active particles and fractions may also be filtered from fluids using ultrafilters. This process is used extensively by beer and wine manufacturers to provide cold stabilization and sterilization of their products. It is also used in water pollution analysis to concentrate organisms from water samples.

Food concentration applications can be applied to processing milk, egg white, animal blood, animal tissue, gelatin and glue, fish protein, vegetable extracts, juices and beverages, pectin solutions, sugar, starch, single-cell proteins, and enzymes. These areas represent a large potential growth market for UF.

Ultracentrifugation

Ultracentrifuges utilize intense gravity forces generated by the centrifugal forces of rapid rotation. These high-g forces are used to determine sedimentation coefficients and to separate subcellular fractions such as protein which may be relatively close in mass.

The ultracentrifuge has been an invaluable research tool in the differential separation of particulate suspensions and in the study of sedimentation rates. It is utilized extensively in the fields of virology and cancer research to separate viral particles from solutions. The introduction of moving-gradient centrifugation permits large volumes of virus-containing fluids to be processed, resulting in all the virus particles settling into a narrow band. This is particularly useful in the preparation of vaccines.

Zonal centrifuge rotors have a variety of shapes and descriptions and can be classified in several ways. The simplest distinction is to call those rotors which hold a fixed volume of sample as *batch type* and those rotors in which the sample can be either a fixed volume or continuously varied as *continuous sample flow with isopycnic banding*, or *flo-band type*. While most batch-type zonal rotors can be operated in existing preparative-type centrifuges, the newer flo-band rotors have a different shape and configuration and require special centrifuge drive systems.

Actual uses of UC include purification and separation of viruses, testing of semiconductors, and protein separation in the manufacture of cosmetics, as well as protein separation from human and animal serum. Potential uses include protein or viral and bacterial separations in special food applications, and the analysis of water samples for biomass, pesticide contamination, and minerals.

Several problems have tended to limit the usefulness of the ultracentrifuge. Ultracentrifuge cells are subjected to large forces and have been prone to problems of leaking. Rotors and bearings deteriorate, requiring that the ultracentrifuge be used at less than its rated maximum rotational speed. Large stresses and forces limit the useful lifetime of the ultracentrifuge. Ongoing research programs dealing with high-strength materials, oil-powered drives, and high-pressure seals are attempting to alleviate these problems. They will have to be solved before the ultracentrifuge will be extensively utilized by industry. Turbines have had to be exchanged every 2,500 operating hours, although the entire unit itself may last for many years. The greatest factors holding back ultracentrifugation from wide use in industry in general are that it is applicable to relatively low flow-rate processes, and it is energy and equipment intensive.

ELECTRODIALYSIS

The principle of electrodialysis (ED) is that electrical potential gradients will make charged molecules diffuse in a given medium at rates far greater than obtained by chemical potentials between two liquids, as in conventional dialysis. When a DC electric current is transmitted through a saline solution, most salts and minerals are dissolved in water as positively charged particles (anions, for example, Na +) and negatively charged particles (anions, for example, Cl^-). The cations migrate toward the negative terminal, or cathode, and the anions toward the positive terminal, the anode. By adjusting the potential between the terminals or plates and the electric current, the flow of ions transported between the plates can be varied.

Added control of the movement of the ions can be obtained by placing sheet-type membranes of cation- or anion-exchange material between the outer plates, as shown diagrammatically in Figure 1.1. These sheets of cation-selective resins and anion-selective resins permit the passage of the respective ions in the solution. Under an applied DC field, the cations and anions will collect on one side of each membrane through which they are transported and vacate the other side. Thus, if a NaCl solution is supplied to the central zone of the cell shown in Figure 1.1, the Na+ ions will migrate through Membrane A, depleting the central zone (termed the *diluting* or *product feed stream*) of the salt ions. The two outer zones where the ions collect are commonly known as the *concentrating* or *brine streams*.

Electrodialysis is applicable to continuous-flow type operations. Multimembrane stacks can be built from alternately spacing anionic- and cationic-selective membranes. Flow of solutions through specific compartments and appropriate recombination of transported ions permit desired enrichment of one stream and depletion of another. A schematic view of a typical stack based on the concept of alternating these concentrating and diluting compartments is shown in Figure 1.2. The feed stream enters each compartment along the top of the figure and flows downward toward the lower exit ports and manifolds. However, as the ionized streams move tangentially along the membranes, cations are transported, or attempt movement, toward the left and anions to the right, causing an alternate build-up and a depletion of ions in adjoining compartments. Thus, one resultant output stream is a diluted product water and the other is a concentrated stream of dissolved salt.

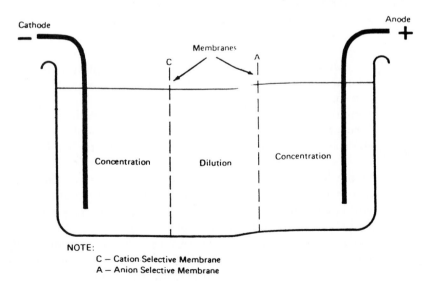

NOTE:
C — Cation Selective Membrane
A — Anion Selective Membrane

Figure 1.1 Electrodialysis cell diagram.

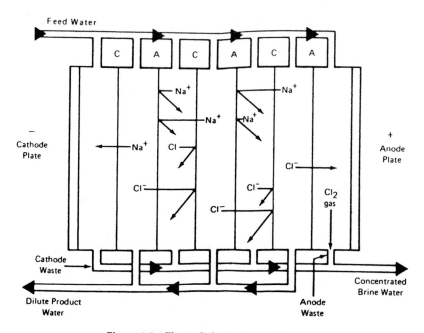

Figure 1.2   Electrodialysis process diagram.

The process flow stream through a commercial demineralizer, incorporating two stacks in series demineralized water, is shown in Figure 1.3. Several of the refinements required for continuous-flow operational systems are shown on this diagram, representing a two-stage demineralizer.

Although ED is more complex than other membrane separation processes, the characteristic performance of a cell is, in principle, possible to calculate from a knowledge of ED cell geometry and the electrochemical properties of the membranes and the electrolyte solution.

Another kind of electrodialysis cell configuration, shown in Figure 1.4, is a multiple electrodialysis system consisting of ten-unit cells, in series rather than manifolded in parallel. The feed solution is introduced at four points: It enters at both upper end points to sweep directly through both electrode chambers and is introduced into the working chambers near either end. The feed solution into the left side traverses depleted chambers and exits as depleted effluent at the right. The feed solution into the rightmost enriching cell flows in the other direction and exits as enriched effluent at the left side.

In addition to the membrane stacks in electrodialysis units, various supporting equipment is essential. This includes pumps for circulation of concentrating and diluting flows; flushing streams for cathode and anode plates; injection systems for pH control; pressure concentration, pH alarms, and control systems and backflushing controls; feed strainers and filters; and grounding systems. Because of the high pH of the cathode stream, substances

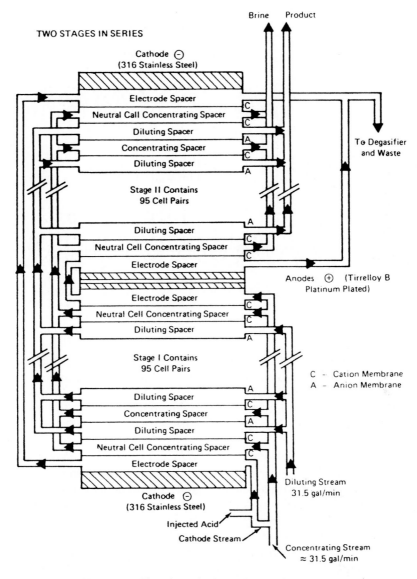

Figure 1.3 Flow through electrodialysis stack.

such as carbonates and hydroxides could precipitate on the cathode surface and adjoining membrane; often sulfuric acid is injected to maintain the stream at pH of 2 or less. Also, recirculating concentrate requires an acid addition to yield a low pH for stability along with additional substances such as sodium hexametaphosphate.

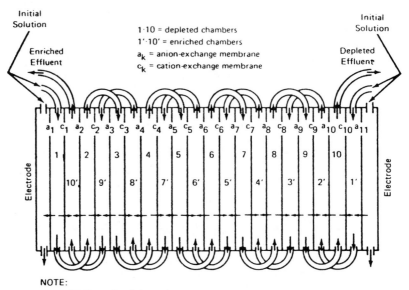

Figure 1.4 Multiple-membrane electrodialysis unit.

Among the problems associated with the electrodialysis process, concentration-polarization is the most significant. This problem which also exists in reverse osmosis systems is because of a build-up in the concentration of ions on one side of the membrane and a decrease in concentration on the opposite side. This adversely affects the operation of membranes and can even damage or destroy them. Polarization occurs when the movement of ions through the membrane is greater than the convective and diffusional movements of ions in the bulk solutions toward and away from the membrane. Along with a deleterious pH shift occurring at the membrane surface, polarization may cause solution contamination and sharply decrease energy efficiency. Commercial electrodialyzer designs incorporate various baffles or turbulence promoters and limit current densities to avoid these effects. Increased feed flow also assists mixing but requires additional power for pumping.

by inorganics in the feed solution and membrane fouling by organics. Efficient separation of suspended matter from the influent stream, for example, by activated carbon absorption, can reduce or prevent such problems. Also, periodic flushing of membranes with acid solutions and/or detergent washing solutions is a conventional practice. An operational means of reducing the scaling and fouling problems consists of cyclical reversal of cathodes and anodes along with interchanging the concentration and dilution streams. An ED variant termed the transport-depletion process, which requires special anion-selective membranes, is also a promising antifouling concept.

A problem with electrodialyzers is influent variability, the remedy for which is efficient monitoring of quality and refined control of pretreatment. Selective ion-permeable membranes are sensitive to specific input charges, including their pH and their concentration; the anion-permeable membranes are particularly susceptible to damage if operated outside their design range.

Applications

Competitive uses of electrodialysis thus far have been in applications in which low concentrations of ions, generally less than 5,000 parts per million (ppm) of total dissolved solids (TDS), are to be removed from solutions. For desalting sea water (which contains about 33,000 ppm of TDS) or for treating brackish industrial wastewaters which are dirty or complex, only evaporative or freezing processes are currently considered commercially viable. The larger-capacity plants, typical of all separation processes, are most attractive from the standpoint of minimum total plant and operational cost per unit of resultant product. Electrodialysis as well as reverse osmosis (RO), ultrafiltration (UF), and ultracentrifugation (UC) have primary applications in:

- Recovery of materials from liquid effluents, that is, processes related to conservation, cleanup, concentration, and separation of desirable fractions from undesirable ones.
- Purification of water sources.
- Effluent water renovation for reuse or to meet point source disposal standards (required to maintain suitable water quality in the receptor streams).

Treatment of brackish waters in the production of potable supplies is the largest present application of electrodialysis procedures. Brackish water sources include both underground supplies and low-quality surface streams or ponds, containing generally less than 5,000 ppm of total dissolved solids. Levels of treatment are customarily achieved which produce an effluent meeting standards of less than 500 ppm of TDS. Moreover, some ED water treatment plants produce high-quality boiler feed water containing only 5 to 10 ppm TDS.

The second largest use of electrodialysis in the United States is in the dairy industry in the desalting of cheese whey. The supernatant liquor associated with the production of cheese contains large amounts of protein, along with lactose, salt, and some acid. Demineralizing the whey—that is, removing the salt and acid by ED—produces a highly marketable infant formula composed of 45 percent demineralized whey.

Electrodialysis has been under study for use in the dairy industry in applications other than whey desalination. These include control of the cation balance in milk, particularly for persons requiring low-sodium intake, and the replacement of strontium by calcium to reduce the radioactive elements in milk and associated products. Both applications have been shown to be technically feasible but have not been brought into commercial use.

An increasing application of ED is the sweetening of prepared citrus juices. In this process, the juice is deacidified through an anion-exchange membrane which removes the citrate ion without degradation of the basic flavor of the product. Although such products can be sweetened by blending or by the addition of sugar, the final product cannot be marketed as a premium product. Anion-exchange membranes were specifically designed for this application since some of the citrus juice components are highly reactive. Because of the large amount of pulp in citrus juice, cell width is increased. The membrane is supported instead by operating the juice compartments at slightly higher pressures than the caustic compartments. Loss in current efficiency because of the build-up of juice constituents in membrane surfaces is overcome by cleaning the surfaces as a result of reversing the current flow at frequent intervals.

Less extensive, industrial uses of electrodialysis currently in commercial operation include the tertiary or advanced treatment of municipal sewage water or of industrial wastewater to remove undesirable ions subsequent to primary and secondary treatments. Other industrial discharges treatable by ED include metal-plating baths, metal-finishing rinse waters, battery manufacturing wastes, wood-pulp wash water, glass-etching solutions, and effluents from desalination processes.

A substantial disposal problem in the metal-finishing industry is due to depleted plating baths and rinse waters. The salt-concerning ability of electrodialysis can be put to use in reconcentrating these metal-plating baths. Materials such as cyanides of zinc and cadmium can be reconcentrated to bath strength and can be reduced in concentration to very low levels. This approach allows recovery of metals while minimizing disposal problems.

Potential Applications

In the ED developmental area, thin membranes suitable for elevated temperature operation to 180°F have been produced. Should these provide lower-cost desalination of high-salinity waters, they might produce potable water from sea water competitively. High-temperature ED advantageously lowers stack energy or membrane surface area requirements.

Electrolyte removal in polymer processing should compete with ion-exchange methods and is preferred for higher concentrations of electrolytes in the feed and when no need is specified for their complete removal.

Bleaching wood pulp with chlorine or hypochlorite solution yields a copious effluent stream of salty water, the disposal of which has become a significant problem in the paper industry. By the use of electrodialysis, an effluent of 4,000 ppm NaCl can be separated into a water stream containing 500 ppm or less of salt and a brine stream of up to 150,000 ppm. The water stream, demineralized to the purity required by the process, is recycled as wash water. The brine stream is electrolyzed in a membrane cell to sodium hydroxide solution and chlorine; these substances may be used directly or part of each may be combined to form sodium hypochlorite solution. Thus, the brine stream is also returned to the process. A very nearly closed-cycle operation is obtained.

Electrodialysis is particularly desirable for the salt-concentration step, since process costs are relatively insensitive to the concentration of the effluent brine. A particular advantage of the electrodialysis process is its ability to produce solutions of high concentrations of soluble salts. A combination of electrodialysis with conventional evaporation may be substantially cheaper than evaporation alone for the production of dry salt from saline solutions.

REVERSE OSMOSIS

When pure water and a salt solution are introduced on opposite sides of a semipermeable membrane in a vented container, the pure water diffuses through the membrane and dilutes the salt solution. At equilibrium, the liquid level on the saline water side of the membrane will be above that on the fresh-water side; this process is known as osmosis and is depicted in Figure 1.5. The view on the left illustrates the commencement of osmosis and the center view presents conditions at equilibrium.

The effective driving force responsible for the flow is osmotic pressure. This pressure has a magnitude dependent on membrane characteristics, water temperature, and salt solution properties and concentration. By applying

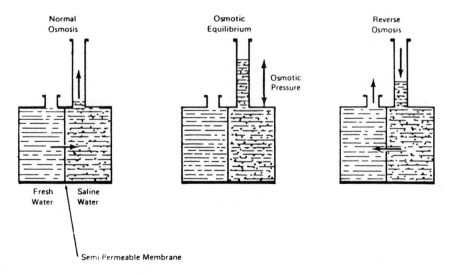

Figure 1.5 Principle of reverse osmosis.

pressure to the saline water, the flow process through the membrane can be reversed. When the applied pressure on the salt solution is greater than the osmotic pressure, fresh water diffuses in the opposite direction through the membrane and pure solvent is extracted from the mixed solution; this process is termed reverse osmosis (RO). The fundamental difference between reverse osmosis and electrodialysis is that in reverse osmosis the solvent permeates the membrane, while in electrodialysis the solute moves through the membrane.

Reverse osmosis is a means for separating dissolved solids from water molecules in aqueous solutions as a result of the membranes being composed of special polymers which allow water molecules to pass through while holding back most other types of molecules; since true "pores" do not exist in the membrane, suspended solids are also retained by *superfiltration*. In an actual reverse osmosis system, operating in a continuous-flow process, feed water to be treated or desalinated is circulated through an input passage of the cell, separated from the output product water passageway by the membrane. The feed stream is split into two fractions—a purified portion called the product water (or permeate) and a smaller portion called the concentrate, containing most of the impurities in the feed stream. At the far end of the feed-water passage, the concentration (dewatered) reject stream exits from the cell. After permeating the membrane, the product (fresh-water) flow is collected. The percentage of product water obtained from the feed stream is termed the *recovery*, typically around 75 percent.

The ratio $(F-P)/F$, or the concentration of a solute species in the feed (F) minus that in the product (P) over the concentration in the feed, is called the *rejection* of that species. Rejections may be stated for particular ions, molecules, or conglomerates such as TDS or hardness. Solids rejection depends on factors such as types and forms of solids, membrane types, recovery, pressure, and pH. Suspended solids (typically defined as particles larger than 0.5 micron mean diameter, and including colloids, bacteria, and algae) are rejected 100 percent; that is, none can pass through the membrane. Weakly ionized dissolved solids (usually organics, but may include other materials such as silicates) undergo about a 90 percent rejection at normal recoveries for certain membranes. Although pH can strongly influence the rejection, when the molecular weight of these solids is less than 100, rejection decreases appreciably. Ionized solids, or salts, are rejected independent of molecular weight and at molecular weights considerably below 100. At 75 percent recovery and pressures greater than 250 pounds per square inch, overall rejection of total dissolved solids (TDS) is about 90 percent. Rejections vary with pressure because the actual salt flow through the membrane remains fairly constant, but the water permeation depends nearly linearly on pressure, affecting the ratio of concentrations. For example, rejection of sodium chloride can fall from 90 percent at 300 pounds per square inch to 20 percent at 50 pounds per square inch, indicating the need to operate at the highest pressures possible.

Cellulose acetate is a common membrane material, but others include nylon and aromatic polyamides. The mechanism at the membrane surface involves the influent water and impurities attempting to pass through the pressurized side, but only pure water and certain impurities soluble in the membrane emerge from the opposite side.

Various configurations of membranes with different surface-to-volume ratios and different flux capabilities (gallons per day per square foot, or gpd/ft²) have been developed. Each type of membrane is a flexible plastic film—no more than 4 to 6 mils thick, firmly supported. Basic designs include the plate and frame, the spiral-wound module (jellyroll configuration), the tubular, and the newest of the process designs, the hollow-fine fiber. Fibers range from 25 to 250 microns (0.001 to 0.01 in.) in diameter, can withstand enormous pressure, are self-supporting, and can be bundled very compactly within a containment pipe. While product flow per square foot of fiber surface is less than that for an equivalent area of flat membrane, the difference in surface area more than compensates for the reduced unit flux.

Major problems inherent in general applications of RO systems have to do with (1) the presence of particulate and colloidal matter in feed water, (2) precipitation of soluble salts, and (3) physical and chemical makeup of the feed water. All RO membranes can become clogged, some more readily than others. This problem is most severe for spiral-wound and hollow-fiber modules, especially when submicron and colloidal particles enter the unit (larger particulate matter can be easily removed by standard filtration methods). A similar problem is the occurrence of concentration-polarization, previously discussed for ED processes. Concentration-polarization is caused by an accumulation of solute on or near the membrane surface and results in lower flux and reduced salt rejection.

The degree of concentration that can be achieved by RO may be limited by the precipitation of soluble salts and the resultant scaling of membranes. The most troublesome precipitate is calcium sulfate. The addition of polyphosphates to the influent will inhibit calcium sulfate scale formation, however, and precipitation of many of the other salts, such as calcium carbonate, can be prevented by pretreating the feed either with acid or zeolite softeners, depending on the membrane material.

Hydrolysis of cellulose acetate membranes is another operational problem and occurs whenever the feed is too acid or alkaline; that is, the pH deviates beyond designed range limits. As may readily happen, whenever CO_2 passes through the membrane, the resultant permeate has a low pH. The operational solution is to remove the gas from the permeate by deaerators, by strong-base anion resins or a complementary system—for example, RO and ion exchange, in series. Aromatic polyamide or nylon membranes are much less sensitive to pH than cellulose acetate.

Compounds such as phenols and free chlorine that are either soluble in the membrane or vice versa will be poorly rejected and may damage the membrane. Procedures to improve feed-water makeup and thus reduce such membrane damage include acid pretreatment of the feed water, dechlorination, periodic cleaning or replacement of the membrane, sequestration of cations, coagulation and filtration of organics, and use of alternative, more durable membrane materials.

Applications

Reverse osmosis process is applied—or undergoing evaluation for imminent application—to a number of water-upgrading needs including high-purity rinse water production for the electronics industry (semiconductor manufacturing), potable municipal water supplied for newly-developed communities (for example, large coastal plants to upgrade brackish well water contaminated by seawater intrusion), boiler feed-water supplies, spent liquor processing for pulp and paper mills, and treatment of acid mine drainage.

In desalting operations, distillation plants have provided the major portion of the world's capacity. As the world's requirements for treated water increase, however, and water quality standards become more stringent, the membrane treatment processes in general and commercial RO processes in particular have been undergoing appreciable development. Important factors in the expansion of commercial RO applications are their favorably low power requirements and the realization of continuous technical improvements in membranes which are used in RO systems. A general guideline in water benefication is that RO is most frequently considered for cases in which the TDS is greater than 2,000 to 3,000 ppm; ED generally applies when the TDS is less than 2,000 to 3,000 ppm. However, many exceptions exist, based on feed-water species and product requirements.

One of the most important applications of RO is in the reclamation of large volumes of municipal and industrial wastewaters and the concentration of the solids for simplified disposal. The value of the reclaimed water offsets the cost of RO, and dilute wastewater concentration leads to economies in any further required liquid waste treatment.

Unrestricted use of reclaimed wastewater for drinking water, however, requires careful examination. While practically a complete barrier to viruses, bacteria, and other toxic entities that must be kept out of a potable supply, RO membranes could pose serious problems should any defect develop in their separation mechanism. Given the purity and clarity of RO-treated wastewaters, however, it might be advantageous to use RO and then subject the product to well-established disinfection procedures.

Smaller-capacity operations include:

- Treatment of industrial wastes:
  - Separation of heavy metal ions, acids, bases, and cyanides from metal-finishing effluents, for example, from alkaline copper baths or zinc cyanide plating rinses.

- Treatment of wastewater from fish and shellfish processing plants and the retrieval of protein substances.
- Removal of toxic compounds from wastewater.
- Processing of pulp and paper mill effluents for reuse.
- Treatment of refinery and petrochemical industry wastewaters, including normal process/utility effluents, ballast water discharge, and contaminated storm runoff.
- Recovery of a medium-quality water from wastewaters in electric generating processes, including air heater wash water, boiler blowdown, and demineralizer regenerant wastes.
- Treatment of laundry effluents.
- Treatment of marine wastes.

Potential Applications

Because of the diversity of water-upgrading needs, there exists a wide range of potential applications for reverse osmosis. Success has been obtained in tests on treatment of irrigation return flows.

Another major potential application for reverse osmosis is sea-water desalination. Feasibility of single-stage high-pressure RO equipment or lower-pressure multiple-stage units might lead to improved operation and costs. Recovery of fresh water from geothermal sources could be another use of RO. Development of new energy technologies may provide several markets for RO processes. Introduction of advanced high-efficiency gas turbines (high-temperature units) in future electrical power generation may require water injection for several reasons, including blade cooling and zonal combustion cooling to inhibit nitrogen oxide formation. Since required water would be of condensate quality to avoid undesirable deposits forming in the turbine or the water passages, extensive development of pure-water production plants would be needed for such turbine installations. Treatment of brackish wastewater from the terminus of slurry pipelines (coal transport) prior to use for irrigation is another prospective application.

Food processing is an area with considerable potential for reverse osmosis. Liquid foods (for example, fruit juices, egg white, milk, coffee) generally contain 80 percent to 99 percent water in their natural state. To an industry concerned with economical preservation, storage, and shipment of food, this water has principally a nuisance value, adding bulk and weight without contributing to nutritional quality. Consequently, the food industry finds it attractive to concentrate liquid foods, removing part of the water by evaporation or freeze-concentration, and occasionally resorting to complete dehydration. Many of these concentrated products have received wide consumer acceptance. Tomato paste, catsup, evaporated milk, soups, and fruit juices are examples of concentrated liquid foods that are in common household use.

Liquid foods are very subjective with respect to flavor and aroma. In many instances their characteristic flavor is due to the presence of minute amounts of volatile substances (or possibly to the action of enzymes on flavor precursors). They are rendered susceptible to damage either by evaporation itself or by the effects of high temperature. While some damage can be compensated for by the addition of essences or flavor concentrates to the product, a better alternative would be to eliminate evaporation and high temperatures altogether. Reverse osmosis and other membrane processes provide a method of concentrating and processing foods economically, without detrimental heating or phase change. Some of the more promising future applications of RO involve the concentration of complex solutions of organic molecules. An example of this is protein retention in the concentration of fish processing wastes.

The concentration-polarization effect is aggravated by the relatively high viscosity of many foods. A prime example of this is the RO concentration of tomato juice. When tomato juice is concentrated at 1,000 psi, dilute juice with an apparent viscosity of only 3 centipoise (cp) flows easily down the center of the channel leaving a thick paste of over 2,000 cp at the membrane surface. (For reference, water at 25°C is about 0.9 cp.) The concentration of the solutes in the juice is only 5 percent at the center of the channel, but 33 percent at the membrane. Flux can be reduced to almost zero in this manner. Grape juice, apple juice, and sucrose solution also present similar problems of solute build-up at the membrane.

A related effect occurs in concentrating lemon juice. In this case, an oil phase might separate out of the juice and "blind" the membrane, greatly reducing flux.

Another potential application is the concentration of caffeine by RO as an alternative to the currently used evaporation procedure.

Reverse osmosis could be used in the food industry in the treatment of wastes and wastewater. In the dairy processing industry, for example, RO could be used for complete recycling of wastewaters (tertiary treatment), a treatment which would include the following sequence of processes: (1) lime precipitation clarification, (2) ammonia stripping, (3) recarbonation, (4) sand filtration, (5) reverse osmosis, and (6) activated carbon.

ULTRAFILTRATION

In ultrafiltration (UF), a feed emulsion is introduced into and pumped through a membrane unit (Figure 1.6). Water and some dissolved low molecular weight materials pass through the membrane under an applied hydrostatic pressure. Emulsified oil droplets and suspended particles are retained, concentrated, and removed continuously as a fluid concentrate. In contrast to ordinary filtration, there is no build-up of retained materials on the membrane filter.

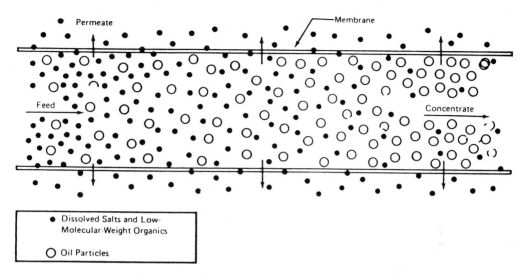

Figure 1.6 Simplified ultrafiltration flow schematic.

The pore structure of the membrane acts as a filter, passing small solutes such as salts, while retaining larger emulsified and suspended matter. The pores of ultrafiltration membranes are much smaller than the particles rejected, and particles cannot enter the membrane structure. As a result, the pores cannot become plugged. Pore structure and size (less than 0.005 microns) of ultrafiltration membranes are quite different from those of ordinary filters in which pore plugging results in drastically reduced filtration rates and requires frequent backflushing or some other regeneration step.

In addition to pore size, another important design criteria in ultrafiltration is the membrane capacity. This is termed *flux* and it is the volume of water permeated per unit membrane area per unit time. The standard units are gallons per day per square foot (gpd/ft^2) or cubic meters per day per square meter (m^3/day/m^2). Because membrane equipment, capital costs, and operating costs increase with the membrane area required, it is highly desirable to maximize membrane flux. A process flow schematic is shown in Figure 1.7.

Ultrafiltration utilizes membrane filters with small pore sizes ranging from 0.015μ to 8μ in order to collect small particles, to separate small particle sizes, or to obtain particle-free solutions for a variety of applications. Membrane filters are characterized by a smallness and uniformity of pore size difficult to achieve with cellulosic filters. They are further characterized by thinness, strength, flexibility, low absorption and adsorption, and a flat surface texture. These properties are useful for a variety of analytical procedures. In the analytical laboratory, ultrafiltration is especially useful for gravimetric analysis, optical microscopy, and X-ray fluorescence studies.

All particles larger than the actual pore size of a membrane filter are

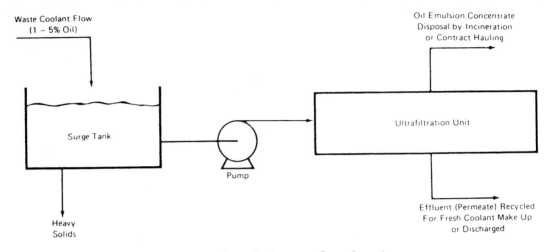

Figure 1.7 Generalized process flow schematic.

captured by filtration on the membrane surface. This absolute surface retention makes it possible to determine the amount and type of particles in either liquids or gases—quantitatively by weight or qualitatively by analysis. Since there are no tortuous paths in the membrane to entrap particle sizes smaller than the pore size, particles can be separated into various size ranges by serial filtration through membranes with successively smaller pore sizes. Figure 1.8 shows pore size in relation to commonly known particle sizes.

Fluids and gases may be cleaned by passing them through a membrane filter with a pore size small enough to prevent passage of contaminants. This capability is especially useful in a variety of process industries which require cleaning or sterilization of fluids and gases. The retention efficiency of membranes is dependent on particle size and concentration, pore size and length, porosity, and flow rate. Large particles that are smaller than the pore size have sufficient inertial mass to be captured by inertial impaction.

In liquids the same mechanisms are at work. Increased velocity, however, diminishes the effects of inertial impaction and diffusion. With interception being the primary retention mechanism, conditions are more favorable for fractionating particles in liquid suspension.

In contrast to reverse osmosis, where cellulose acetate has occupied a dominant position, a variety of synthetic polymers has been employed for ultrafiltration membranes. Many of these membranes can be handled dry, have superior organic solvent resistance, and are less sensitive to temperature and pH than cellulose acetate. Polycarbonate resins, substituted olefins, and polyelectrolyte complexes have been employed among other polymers to form ultrafiltration membranes. Preparation details for most of the membranes, however, are proprietary. Molecular weight cutoff is used as a measure of rejection. However, shape, size, and flexibility are also important parameters. For a given molecular weight, more rigid molecules are better rejected than

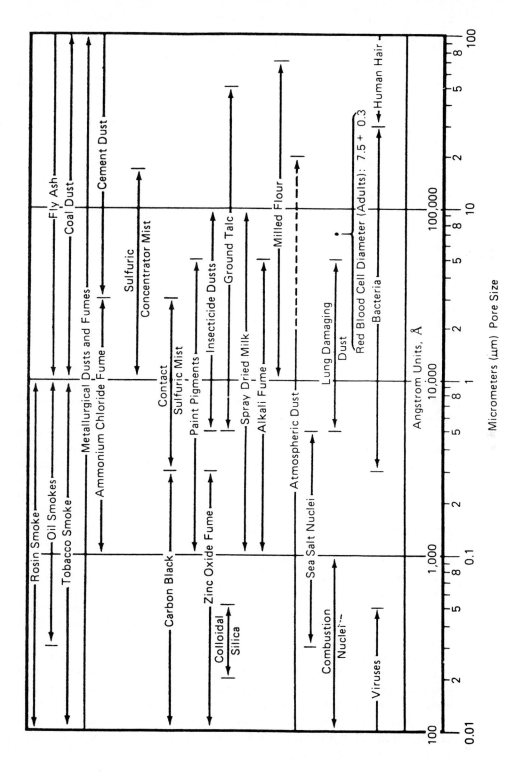

Figure 1.8 Ranges of common particle sizes (diameter) over range of UF pore size.

21

flexible ones. Ionic strength and pH often help determine the shape and rigidness of large molecules.

Operating temperatures for membranes can be correlated generally with molecular weight (mol wt) cutoff. For example, maximum operating temperatures for membranes with 5,000 to 10,000 mol wt cutoffs are about 65°C. For a 50,000 to 80,000 mol wt cutoff, maximum operating temperatures are in the range of 50°C.

Membrane lifetimes are usually two years or more for treating clean streams (water processing), but are drastically reduced when treating comparatively dirty streams (oil emulsions). Membrane guarantees by manufacturers are determined only after pilot work is done on the particular stream in question. In some cases as little as a 90-day guarantee may be given for oil/water waste applications.

Applications

The largest industrial use of ultrafiltration is the recovery of paint from water-soluble coat bases (primers) applied by the wet electrodeposition process (electrocoating) in auto and appliance factories.

The next largest application is the recovery of proteins in cheese whey (a waste from cheese processing) for dairy applications. This is done if a market for protein can be found, in particular for feeding cattle and farm animals. In cheese whey processing, a typical unit might process 500,000 pounds a day of whey for 300 days a year.

Another large-scale application is the concentration of waste-oil emulsions from machine shops, which are produced in association with cooling, lubricating, machining, and heavy metal rolling operations. The separation of the oil from the water works well with stable emulsions, but with unstable emulsions the oil will clog the filter.

Ultrafiltration is a preferred alternative to the conventional systems of chemical flocculation and coagulation followed by dissolved air flotation. Ultrafiltration provides lower capital equipment, installation, and operating costs.

Biologically active particles and fractions may be filtered from fluids using ultrafilters. This process is used extensively by beer and wine manufacturers to provide cold stabilization and sterilization of their products. It is also used in water pollution analysis to concentrate organisms from water samples.

Small-pore filters $(0.015–0.1\mu)$ may be used to filter viruses, and proteins may be filtered out using filters of 0.05μ or less. Current uses include concentration of proteins in blood plasma as a replacement for dialysis of blood and for producing colloid-free water for cosmetics.

Ultrafilters are capable of filtering cells and cell fractions from fluid media. These particles, after concentration by filtration, may be examined through subsequent quantitative or qualitative analysis. The filtration techniques also have applications in fields related to immunology and implantation of tissues as well as in cytological evaluation of cerebrospinal fluid.

Potential Uses

Current uses of ultrafilters cover extensive areas of biological research, processing sterile fluids, air and water pollution analysis, and recovery of corrosive or noncorrosive chemicals through filtration. Future uses of ultrafilters are likely to be logical extensions of current uses, such as use by food industries to produce sterile liquids without heat treating, filtering out noxious pollutants from process plant waste fluids, and extensive use by medical laboratories and pharmaceutical houses to concentrate or filter out viruses, cells, cell fragments, or proteins.

Ultrafiltration may be applicable to dewatering some sludges, but this use is highly dependent on the particular sludge itself. There are no commercial uses of UF for sludge dewatering at this time, but several sources have been found which claim that this represents a possible future application.

Pollution of water supplies within the food industry is a significant problem, since many food wastes possess extremely high biochemical oxygen demand (BOD) requirement. In the potato starch industry, for example, waste effluent containing valuable proteins, free amino acids, organic acids, and sugars can be processed by ultrafiltration. Reclamation of these materials, which are highly resistant to biodegradation, could provide an economic solution to this waste removal problem.

For concentration of juices and beverages, RO is preferable to evaporation due to lower operating costs and no degradation of the product. Processing by RO retains more flavor components than does heated, vacuum-pan concentration. Since ultrafiltration does not retain the low molecular weight flavor components and some sugar, UF is employed as a complement to RO. A two-stage process may be used in which the first stage, UF, allows the passage of sugars and other low molecular weight compounds. This permeate is then dewatered by RO and recycled back to the main stream. High juice concentrations are possible in this manner because the UF removes the colloidal and suspended solids which would foul the RO, and helps relieve some of the high hydraulic pressure due to high osmotic pressure of the juice.

In a process as shown in Figure 1.9, a citrus press liquor, or multiphase suspension, is ultrafiltered following a coarse prefiltration. The resultant clear permeate is processed through ion exchange and granular activated carbon adsorption units to remove low molecular weight contaminants and inorganic salts. The product is a natural citrus sugar solution suitable for reuse. The concentrated suspended solids are used in making animal feed.

Pectins are a family of complex carbohydrates which are used to form gels with sugar and acid in the production of jellies, preserves, and other confections from fruit juices.

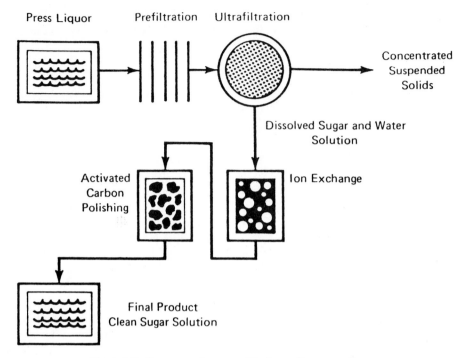

Press Liquor   Prefiltration   Ultrafiltration

Concentrated
Suspended
Solids

Dissolved Sugar and Water
Solution

Activated
Carbon
Polishing

Ion Exchange

Final Product
Clean Sugar Solution

Figure 1.9 Recovery of sugar solids from citrus press liquor.

The recovery of starch and other high molecular weight compounds from waste effluents is an important application for UF. The output from a 30-ton-per-day starch plant would be about 432,000 gpd with a solids content of 0.5 percent to 1.0 percent, and 9,000 to 14,000 mg/l COD. Treatment systems which can both reduce the strength of this waste and recover valuable by-products, such as proteins, are an obvious advantage to this industry.

Heat and acid coagulation, distillation, and freezing techniques are more costly and less efficient than UF in protein recovery. Reverse osmosis is also a competing process, but protein recovery by UF would lead to a somewhat higher purity.

In the production of single-cell proteins as a food source, UF has several applications. For harvesting cells, UF could replace centrifugation since the efficiency of centrifugation decreases rapidly with particle size. Ultrafiltration is also well suited for recovering and concentrating the metabolic products of fermentation (enzymes, for example). In a related application, UF is also able to concentrate and desalt protein products, being more efficient than dialysis for this purpose. Moreover, a UF membrane module may be coupled with a fermenter so that toxic metabolites can be continuously removed from the system as fresh substrate is introduced. This permits the growth limitations of a batch fermenter to be relieved and permits a substantial increase in productivity.

A membrane enzymatic reactor is similar to a membrane fermenter with the exception that no microorganisms are present. Instead, enzyme-catalyzed reactions take place and reuse of the enzyme is simplified. That is, purification problems and enzyme removal from end products can be eliminated. The current thrust of research on ultrafiltration is toward lowering the cost of both the equipment and the filters and extending the range of applications to waste treatment and concentration problems.

ULTRACENTRIFUGATION

The ultracentrifuge utilizes intense gravity forces (up to 400,000 g) generated by the centrifugal forces of a rapidly rotating rotor (up to 70,000 rpm). These high gravity forces are used to determine sedimentation coefficients and to separate subcellular fractions such as protein, which, because they may be relatively close in mass, are difficult to separate by other means.

Centrifugation may be used in the differential separation of heterogeneous particulate suspensions. The application of gravity forces of rotation causes sedimentation of the particulates suspended in the solution, as shown in Figure 1.10.

At the completion of centrifugation, large particles are separated into a *pellet* at the bottom of the centrifuge tube while small particles are still suspended in solution. Generally, only partial separation is initially possible using differential centrifugation since the pellet zone is contaminated with

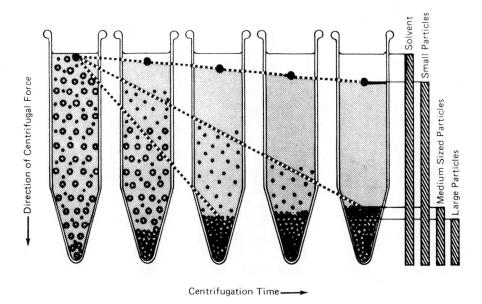

Figure 1.10 Sedimentation of a particulate suspension in a centrifugal field.

211 particle species because all particles will settle at different rates. To solve this problem, rate-zonal centrifugation uses liquid-density gradients, resulting in a much higher resolution as shown in Figure 1.11.

The mass of material which can be separated by rate-zonal centrifugation in a density gradient, however, is severely limited. Other problems associated with rate-zonal centrifugation include the *wall effect*, in which settling particles hit the wall of the centrifuge before they reach the bottom of the tube, and *streaming*, in which the sample layer tends to break up into droplets which settle to the bottom through the gradient.

To provide for maximum sample sizes while minimizing the wall and streaming effects, a hollow bowl-shaped rotor is introduced; the interior is divided into sector-shaped compartments using radially arranged vertical septa. The septa serve to insure uniformity of acceleration and deceleration of the fluid in the rotor, preventing switching or loss of separation resolution. Septa also serve to contain flow lines used to fill or empty rotors. Gradients and samples are generally introduced into the rotor while it is spinning, and recovery of the gradient with its contained separated zones is accomplished in the same manner. The gradient may be moved radially during rotation by pumping dense fluid to the rotor edge or light fluid to the rotor center. This enables combinations of rate and isopycnic separations. For example, cell membranes and mitochondria band isopycnically in the same layer of a su-

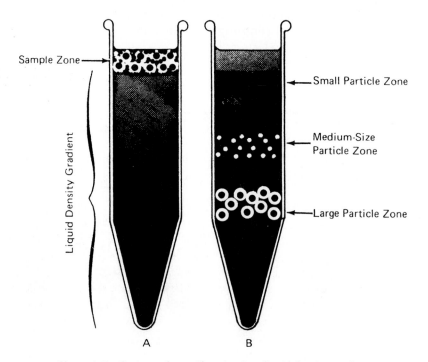

Figure 1.11 Rate-zonal centrifugation in a liquid density gradient.

crose gradient solution. They settle at different rates, however, so the more slowly moving mitochondria may be unloaded with the sample layer while the more rapidly settling cell membranes have banded in the sucrose gradient. While many particles in cell homogenates have sedimentation coefficients in the range of most viruses, few have the same sedimentation rate and banding density as viruses. Thus, this method may be used to separate these particles.

The possibility of moving the gradients during rotation permits large volumes of virus-containing fluids to be introduced inboard to a small steep gradient, thereby settling all the virus into a narrow band. By replacing the sample layer many times, the virus particles from large volumes of sample may be recovered and simultaneously purified by isopycnic banding. The preparation of vaccines particularly has been aided by the development of continuous charging of the sample layer at high speed (continuous sample flow with isopycnic banding).

Zonal centrifuge rotors have a variety of shapes and descriptions and can be classified in several ways. The simplest distinction is to call the rotors which hold a fixed volume of sample as *batch type* and those rotors in which the sample can either be a fixed volume or continuously varied as *continuous sample flow with isopycnic banding*, or *flo-band type*. While most batch-type zonal rotors can be operated in existing preparative-type centrifuges, the newer flo-band rotors have a different shape and configuration and require special centrifuge drive systems. It is this class of centrifuge rotors and systems which are discussed here.

Applications

One application of the ultracentrifuge is in the purification and separation of viruses. The continuous-sample flow-banding rotors are used to purify vaccines and isolate viral fractions. Virus recovery, however, is approximately half that predicted theoretically, due in part to back-mixing and channeled flow.

The large gravity forces produced by the ultracentrifuge are used to determine the integrity and reliability of semiconductors. Special rotors and inserts for ultracentrifuges have been developed for this application. The procedure is not widely used but may become important as semiconductors are used in applications requiring high reliabilities under severe loading.

Ultracentrifuges are used by the cosmetics industry to separate protein fractions from solutions used in the manufacture of cosmetics. Protein fractions are common allergens capable of causing severe-contact allergic reactions. It is desirable, therefore, that solutions used in the manufacture of cosmetics be free of these agents. The high-gravity forces of the ultracentrifuge are utilized to settle out protein particles from commercial solutions to produce nonallergenic cosmetics. The use of ultracentrifugation for this purpose is becoming more common.

Flo-band rotors, which were made by converting batch-type rotors for continuous-sample flow capabilities, can still be used for rate separations. Because of this characteristic, the flo-band ultracentrifuge is used in the separation of classes of proteins from both human and animal serum. The large capacity of the rotors permits the use of large samples, and separations are rapidly made because the sedimentation distance is relatively short.

The ultracentrifuge might be used by special food industries to separate undesirable protein fractions from solutions. Viral and bacterial particles may be separated from certain food products through ultracentrifugation, thereby sterilizing the food product. In very specialized applications, certain protein fractions that are particularly allergenic may be separated from food products to produce allergen-free food products.

These applications are a direct outgrowth of the ability of ultracentrifuge to isolate protein fractions and viral particles from solutions. The technology has not been adopted by the food industry, however, because it represents the introduction of a significantly new process which could require considerable investments in new equipment and training. Also, other methods of separation such as filtration are generally available which are less equipment intensive, do not have severe volume constraints, and are more familiar to the food industry.

Flo-band rotors have been used for the analysis of waterborne particulates of both inorganic and organic origin. Figure 1.12 shows a fractionation scheme used for the analysis of sea water for the determination of the distribution of biomass and its viability, and levels of pesticide contamination. While fresh-water samples can be processed in aluminum rotors, only titanium may be used for sea water because of corrosion problems.

The mineral-prospecting capability of these rotors has not yet been exploited. During water fractionation studies mineral colloids are isolated; a systematic analysis of these colloids permits not only a quantitative determination of the minerals present but also an analysis of radionucleides present in the water.

Ultracentrifugation has proven invaluable in biological research applications involving the separation of protein fractions, cell parts, or viral particles. This capability has been utilized by the medical industries for the production of vaccines and solutions free of these particles. The ultracentrifuge has also been useful for the characterization and analysis of cell fractions and proteins by determining sedimentation rates and coefficients.

In the past, several problems have tended to limit the usefulness of ultracentrifuges. Ultracentrifuge cells are subjected to large gravity forces and have been prone to problems of leaking. Rotors and bearings deteriorate with use, requiring that the ultracentrifuge be used at less than its rated maximum rotational speed. Large stresses and forces limit the useful lifetime of the ultracentrifuge. Ongoing research programs dealing with high-strength materials, oil-powered drives, and high-pressure seals are attempting to alleviate these problems. They will have to be solved before the ultracentrifuge will be extensively utilized by industry. Currently, the turbines have to be ex-

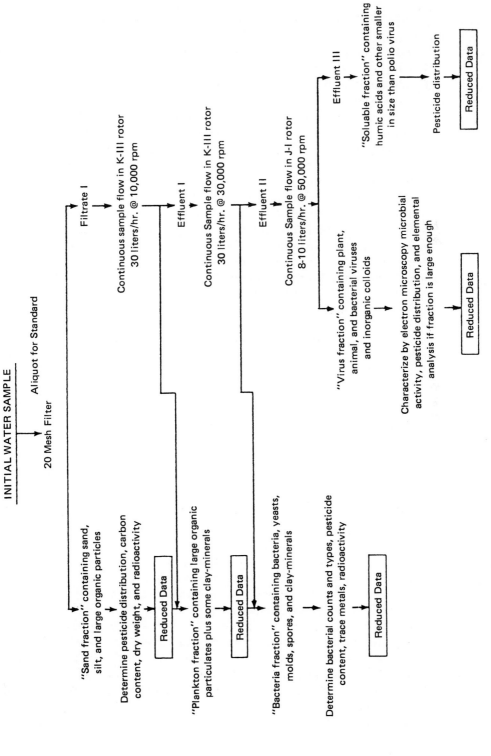

Figure 1.12 A fractionation scheme for the isolation of several classes of waterborne particulates.

changed every 2,500 operating hours, although the entire unit itself may last for many years.

As the resolution of separation methods increases, so does the number of analyses required to evaluate the separation. To deal with this problem, a computer-interfaced fast analytical system was developed which allows many reactions to be done in parallel. This system is finding wide use in the clinical laboratory but has not been employed to evaluate zonal separations.

Future technological developments will include higher-strength materials of construction for greater reliability and less maintenance, very high-speed seals, and new rotor configurations including rotors with more than two fluid lines for sample recovery or gradient reorientation. For virological separations, the overriding problem is that of containment.

The greatest factors holding back ultracentrifugation in general are that it is applicable to relatively low flow-rate processes, and it is energy and equipment intensive. In addition, ultrafiltration methods, which may be used in many cases to concentrate or separate viral particles, are usually much less expensive than ultracentrifugation methods.

2

Chemical Technologies for Water Disinfection

WATERBORNE DISEASES

Untreated waters contain a number of harmful pollutants which give the water color, taste, and odor. These pollutants include viruses, bacteria, organic materials, and soluble inorganic compounds, and these must be removed or rendered harmless before the water can be used again.

A breakdown of the documented outbreaks identifies acute gastroenteritis, hepatitis shigellosis, ciardiasis, chemical poisoning, typhoid fever, and salmonellosis. Sources of contaminated water can be traced to semipublic water systems, municipal water systems, and to individual water systems.

In cell culture, it has been shown that one virion can produce infection. In the human host, because of acquired resistance and a variety of other factors, the one virion/one infection possibility does not exist.

Very little is known of the epidemiology of waterborne diseases. The current database is insufficient to determine the scope and intensity of the problem. The devastating effect of epidemics is sufficient to rank water-associated epidemics as a most important public health problem.

Viruses and bacteria may be eliminated by chemical methods or by irradiation, and organic poisons may also be controlled. Inorganic matter must be removed by other means.

Viruses and Bacteria

Viruses are ultramicroscopic organisms. They are parasites; they need to infest a host in order to duplicate themselves. Viruses excreted with human and animal feces are called enteric viruses, and more than 100 such organisms have been identified. As many as one million viruses can be found in one gram of excrement. The concentration in raw sewage varies over a wide range; as many as 463,500 infectious particles per liter of raw sewage have been detected.

Viruses found in surface waters are introduced from three major sources. Viruses of human origin can be traced to untreated or inadequately treated domestic sewage. Runoffs from agricultural land, feedlots, and forests introduce viruses from domestic and wild animals and birds. Plant viruses, insect viruses, and other forms of life associated with the aquatic environment may also infect the waters.

In addition to viruses, bacteria (microscopic organisms that can reproduce without a host in the proper conditions) are also found in water. In general, damage to the human body from bacterial infection is due to the action of the toxins they produce. Bacteria found in water are derived from contact with air, soil, living and decaying plants and animals, and animal excrements.

Many of these bacteria are aerobic and anaerobic spore-forming organisms associated with varying densities of coliforms, fecal coliforms, fecal streptococci, staphylococci, chromogenic forms, fluorescent strains, nitrifying and denitrifying groups, iron and sulfur bacteria, proteus species, and pathogenic bacteria. Many bacteria are of little sanitary significance and die rapidly in water.

Fecal pollution adds a variety of intestinal pathogens. The most common genera found in water are salmonella, shigella, vibrio, mycobacterium, pasteurella, and leptospira.

The circumstances under which water becomes contaminated are as varied as the ways water is taken internally. It is then conceivable that almost any virus could be transmitted through the water route.

The increased use of water for recreational purposes increases the incidence of human contact with bodies of water and, consequently, with waterborne viruses and bacteria. The major waterborne viruses among pathogens, and the most likely candidates for water transmission, are the picornaviruses (from pico, meaning very small, and RNA, referring to the presence of nucleic acid). The characteristics of picornaviruses are shown in Table 2.1.

Among the picornaviruses are the enteroviruses (polioviruses, coxsackieviruses, and echoviruses) and the rhinoviruses of human origin. Also included are enteroviruses from excrements of cattle, swine, and other domesticated animals; and rhinoviruses of nonhuman origin, viruses of foot and

TABLE 2.1 PICORNAVIRUS CHARACTERISTICS (VERY SMALL RNA VIRUSES)

1. Small spheres about 20–30 μ in diameter
2. RNA core, icosahedral form of cubic symmetry
3. Resistant to ether, chloroform, and bile salts, indicating lack of essential lipids
4. Heat stabilized in presence of divalent cations (Molar $MgCl_2$)
5. Enteroviruses separated from rhinoviruses by acid lability of the latter viruses (inactivated at pH 3.0–5.0)

mouth disease, teschen disease, encephalomyocarditis, mouse encephalomyelitis, avian encephalomyelitis, and vesicular exantherm of pigs.

Additionally, certain viruses can be transported by the water route because their vectors, water molds, and nematodes live in the soil and move with the movement of water. Plant pathogenic viruses also enter the water route and contribute to the problem, though this area has been given little

TABLE 2.2 THE ELECTROMAGNETIC SPECTRUM

Frequency Range	Vacuum Wavelength	Application	Method of Generation
Up to 20 kHz	Down to 10 km	Power and telephone	Rotating generators, microphones, transistors, and tubes
20 kHz–500 MHz	10 km–0.5 m	Radio and TV	Tubes and transistors
555 kHz–1,605 kHz		AM radio	
54 MHz–88 MHz		TV channels 2–6	
88 MHz–108 MHz		FM radio	
174 MHz–46 MHz		TV channels 7–13	
500 MHz–5,000 GHz	15 m–0.1 mm	Industrial radar	Semiconductors, klystrons
470–890 MHz		TV channels 14–82	Magnetrons, masers
5000 GHz–5×10^5 GHz	0.1 mm–0.7 μ	Heat	Lamps, lasers
5×10^5 GHz–3×10^{17} Hz	0.7 μ–0.4 μ	Light	Lamps, lasers
7.5×10^5 GHz–3×10^{17} Hz	0.4 μ–10 Å	UV	UV lamps, spark discharge, lasers, laser-doubling
3×10^{17} Hz–3×10^{20} Hz	10 Å–0.1 Å	X-rays	X-ray tubes, radioactive isotopes
3×10^{20} Hz–3×10^{24} Hz	0.01 Å–0.001 mÅ	Gamma rays	Linear accelerators, nuclear reactions
3×10^{24} Hz–Up	0.001 mÅ and shorter	Cosmic rays	Cosmic sources

attention in the past. Viruses associated with industrial abattoirs, meat packing, food processing, pharmaceutical, and chemical operations are also a potential problem. All enteric viruses occur in sewage in considerable numbers, and recent detection techniques make it possible to find these viruses in almost all streams that receive sewage effluents. Enteric viruses have been isolated from surface waters around the world (Table 2.2). Samples collected from tidal rivers in the United States contained viruses in 27 percent to 52 percent of the cases. The contamination of surface water by enteric viruses appears to be ubiquitous.

A variety of factors (Figure 2.1) is responsible for the survival of viruses in water bodies. The survival of enteric viruses under laboratory conditions and in estuaries varies from a few hours to up to 200 days. Survival in winter is superior to that at summer temperatures.

It is not known exactly what happens to these multitudes of viruses introduced in water bodies. The inability of rhinoviruses to withstand pH changes, temperature fluctuations, and the lack of protective covering offered by feces and other organic materials probably makes the water route of minor importance in their transmission. These factors do not affect the enteroviruses, which are stable and persist in water for long periods. Coxsackieviruses, it has been found, are relatively resistant to concentrations of chlorine normally used for disinfection of bacteria in water. Studies have shown that enteric viruses easily survive present sewage treatment methods and may survive in waters for a considerable time.

Sea or estuary water	2–130 days
River water	2–>188 days
Tap water	5–168 days
Soil	25–175 days
Oysters	6–90 days
Landfill leachates	7–>90 days
Sediments	?

It has been shown that oysters incorporate poliovirus into their tissues even when grown in sea water with only small amounts of virus.

Treatment Methods

The traditional processes and techniques currently in use for the removal of viruses from water and wastewater include methods effecting physical removal of the particles and those causing the inactivation or destruction of the organism. Among the first are sedimentation, adsorption, coagulation and precipitation, and filtration. The second encompasses high pH and chemical oxidation by disinfectants such as halogens.

Primary treatment of municipal waste involving settling and retention removes very few viruses. Sedimentation effects some removal. Virus removal

Required events for transmission
of animal viruses to man by water

Factors influencing virus cycle
and survival

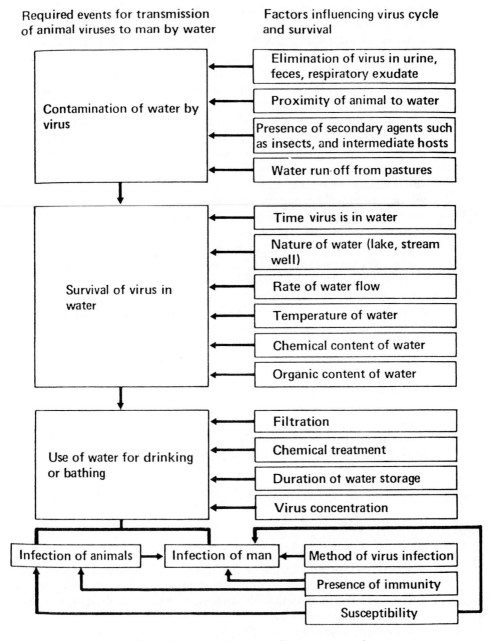

Figure 2.1 Factors that may affect virus survival in water.

of up to 90 percent (which is a minimal removal efficiency) has been observed after the activated sludge step. Further physical-chemical treatment can result in large reductions of virus titer, coagulation being one of the most effective treatments achieving as much as 99.99 percent removal of virus suspended in water. If high pH (above 11) is maintained for long periods of time, 99.9 percent of the viruses can be removed.

Of all the halogens, chlorine at high doses (40 mg/l for 10 min) is very effective, achieving 99.9 percent reduction. Lower doses (for example, 8 mg/l) result in no decrease in virus.

As a result of several studies, the following conclusions regarding viruses in sewage warrant consideration: (1) primary sewage treatment has little effect on enteric viruses; (2) secondary treatment with trickling filters removes only about 40 percent of the enteroviruses; (3) secondary treatment by activated sludge treatment effectively removes 90 percent to 98 percent of the viruses; and (4) chlorination of treated sewage effluents may reduce, but may not eliminate, the number of viruses present.

The current concept of disinfection is that the treatment must destroy or inactivate viruses as well as bacillary pathogens. Under this concept, the use of coliform counting as an indicator of the effectiveness of disinfection is open to severe criticism given that coliform organisms are easier to destroy than viruses by several orders of magnitude.

An important concept is that a single disinfectant may not be capable of purifying water to the desired degree. Also, it might not be practicable or cost effective. This has given rise to a variety of treatment combinations in series or in parallel. The analysis further indicates that the search for the perfect disinfectant for all situations is a sterile exercise.

It has been estimated that in the United States only 60 percent of municipal waste effluent is disinfected prior to discharge and, in a number of cases, only on a seasonal basis. Coupling this fact with the demonstration that various sewage treatment processes achieve only partial removal of viruses leaves us with a substantial problem to resolve.

NONCONVENTIONAL TREATMENT METHODS

Electromagnetic Waves (EM)

Electromagnetic radiation is the propagation of energy through space by means of electric and magnetic fields that vary in time. Electromagnetic radiation may be specified in terms of frequency, vacuum wavelength, or photon energy. The electromagnetic spectrum with its application to water treatment is shown in Table 2.2.

For water purification, EM waves up to the low end of the UV band will result in heating the water. (This includes infrared as well as most lasers.) In the visible range, some photochemical reactions such as dissociation and increased ionization may take place. At the higher frequencies, it will be necessary to have thin layers of water because the radiation will be absorbed in a relatively short distance. It should be noted that the conductivity and dielectric constant of materials are, in general, frequency dependent. In case of the dielectric constant, it decreases as 1/wavelength. Hence, the electromagnetic absorption will vary with the frequency of the applied field. There may be some anomalies in the absorption spectra in the vicinity of frequencies that could excite molecules. At those frequencies, the absorption could be unusually large.

Ultraviolet radiation in the region between 0.2μ to 0.3μ has germicidal properties. The peak germicidal wavelength is around 0.26μ. This short UV is attenuated in air and, hence, the source must be very near the medium to be treated. The medium must be very thin as the UV will be attenuated in the medium as well.

X-rays and gamma rays are high-energy photons and will tend to ionize most anything with which they collide. They could generate UV in air. At higher energies it is possible for the gamma rays to induce nuclear reactions by stripping protons or neutrons from nuclei. This could result in the production of isotopes and/or the production of new atoms.

Sound

A sound wave is an alteration in pressure, stress, particle displacement, particle velocity, or a combination of these that is propagated in an elastic medium. Sound waves, therefore, require a medium for transmission; that is, they may not be transmitted in a vacuum.

The sound spectrum covers all possible frequencies. The average human ear responds to frequencies between 16 Hz and 16 kHz. Frequencies above 20 kHz are called ultrasonic frequencies. Sound waves in the 50–200 kHz range are used for cleaning and degreasing.

In water purification applications, ultrasonic waves have been used to effect disintegration by cavitation and mixing of organic materials. The waves themselves have no germicidal effect but, when used with other treatment methods, can provide the necessary mixing and agitation for effective purification.

Electron Beams

The electron is the lightest stable elementary particle of matter known and carries a unit of negative charge. It is a constituent of all matter and can be found free in space.

ELECTRON CONSTANTS

Mass	$0.109 \times 10{-}31$ kg
Charge	$1.602 \times 10{-}14$ Coulomb
Spin	0.5

Under normal conditions, each chemical element has a nucleus consisting of a number of neutrons and protons, the latter equal in number to the atomic number of the element. Electrons are located in various orbits around the nucleus. The number of electrons is equal to the number of protons, and the atom is electrically neutral when viewed from a distance. The number of electrons that can occupy each orbit is governed by quantum mechanical selection rules. The binding energy between an electron and its nucleus varies with the orbit number, and in general the electrons with the shortest orbit are the most tightly bound. An electron can be made to jump from one orbit into another by giving it a quantum of energy. This energy quantum is fixed for any given transition and whether a transition will occur is again governed by selection rules. In other words, although an electron is given a quantum of energy sufficient to raise it to an adjacent higher state, it will not go up to that state if the transition is not permitted. In that case, it is theorized that if the electron absorbs the quantum, it will most probably go up to the excited state, remain there for a time allowed by the uncertainty principle, reradiate the quantum, and return to its original state.

If an electron is given a sufficiently large quantum of energy, it will completely leave the atom. The electron will carry off as kinetic energy the difference between the input quantum and the energy required to ionize. The remaining atom will now become a positively charged ion, and the stripped electron will become a free electron.

This electron may have sufficient energy when it leaves the atom (or it may acquire sufficient energy from an external field) to collide with another atom and strip it of an electron. This is the basis for electric discharge where free electrons are accelerated by an applied field and, as they collide with neutral atoms, generate additional free electrons. This process avalanches as the electrons approach the positive electrode. At the same time, the positively charged ions are accelerated toward the negative electrode.

In a vacuum, when a voltage is applied between two electrodes, electrons will move from the cathode to the anode. Of course, in a vacuum there will be no avalanching effects. Electrons are emitted from the cathode by a number of mechanisms:

Thermionic Emission. Because of the nonzero temperature of the cathode, free electrons are continuously bouncing inside. Some of these have sufficient energy to overcome the work function of the material and can be found in the vicinity of the surface. The cathode may be heated to increase this emission. Also to enhance this effect, cathodes are usually made of, or coated with, a low work-function material such as thorium.

Shottky Emission. This is also a thermionic type of emission except that in this case, the applied electric field effectively decreases the work function of the material, and more electrons can then escape.

High Field Emission. In this case, the electric field is high enough to narrow the work-function barrier and allow electrons to escape by tunneling through the barrier.

Photoemission. Electromagnetic radiation of energy can cause photoemission of electrons whose maximum energy is equal to or larger than the difference between the photon energy and the work function of the material.

Secondary Emission. Electrons striking the surface of a cathode could cause the release of some electrons and, hence, a net amplification in the number of electrons. This principle is used in the construction of photomultipliers where light photons strike a photoemitting cathode releasing photoelectrons. These electrons are subsequently amplified striking a number of electrodes (called dynodes) before they are finally collected by the anode.

Electromagnetism

In a high-gradient magnetic separator, the force on a magnetized particle depends on the intensity of the magnetizing field and on the gradient of the field.

When a particle is magnetized by an applied magnetic field, the particle develops an equal number of north and south poles. Hence, in a uniform field, a dipolar particle experiences a torque, but not a net tractive force. In order to develop a net tractive force, a field gradient is required; that is, the induced poles at the opposite ends of the particle must view different magnetic fields.

In a simplified, one-dimensional case, the magnetomotive force on a particle is given by

$$F_m = \mu(\delta H/\delta x) = MV(\delta H/\delta x) = \chi HV(\delta H/\delta x)$$

where μ is the magnetic moment of the particle under field intensity, $H\delta H/\delta x$ is the field gradient. The magnetic moment μ is the product of the magnetization of the particle and its volume ($\mu u = MV$). And magnetization is the product of the particle susceptibility, X, and the field intensity, H. In water purification, this magnetic force may be used to separate magnetizable particles.

Direct and Alternating Currents

Electrolytic treatment is achieved when two different metal strips are dipped in water and a direct current is applied from a rectifier. The higher the voltage, the greater the force pushing electrons across the gap between the electrodes. If the water is pure, very few electrons cross the path between the electrodes. Impurities increase conductivity, hence decreasing the required voltage. Additionally, chemical reactions occur at both the cathode and the anode. The major reaction taking place at the cathode is the decomposition of water with the evolution of hydrogen gas. The anode reactions are oxidations by four major means: (1) oxidation of chloride to chlorine and hypochlorite, (2) formation of highly oxidative species such as ozone and peroxides, (3) direct oxidation by the anode, and (4) electrolysis of water to produce oxygen gas.

APPLICATIONS

Electrolytic Treatment

A great deal of interest was generated in the United States prior to 1930 in electrolytic treatment of wastewater, but all plans were abandoned because of high cost and doubtful efficiency. Such systems were based on the production of hypochlorite from existing or added chloride in the wastewater system. A great deal of effort has been made in reevaluating such techniques.

Reduction in Number of Viable Microorganisms by Adsorption onto the Electrodes. Protein and microorganism adsorption on electrodes with anodic potential has been documented. Microorganism adsorption on passive electrodes (in the absence of current) has been observed with subsequent electrochemical oxidation. This does not appear to be a major route for inactivation.

Electrochemical Oxidation of the Microorganism Components at the Anode. Oxidation of various viruses due to oxidation at the surface of the working electrode has been indicated, although the peak voltage used in many experiments would not be sufficient for the generation of molecular or gaseous oxygen.

Destruction of the Microorganisms by Production of a Biocidal Chemical Species. It has been shown that NaCl is not needed for effective operation in the destruction of microorganisms. Biocidal species such as Cl, HO−, O, ClO, and HOCl occur but have very low diffusion coefficients. Hence, if this phenomenon occurs, the probability is that organisms are destroyed at the electrode surface rather than in the bulk solution.

Destruction by Electric Field Effects. It has been observed that some organisms are killed in midstream without contact with the electrodes. The organisms were observed to oscillate in phase with the electric field. Hence, microorganism kill can also be ascribed to changes caused by changing electromotive forces resulting from the impressed AC.

Electromagnetic Separation

In the typical operation, a magnetized fine-particle seed (typically iron oxide) and a flocculent (typically aluminum sulfate) are added to the wastewater, prompting the formation of magnetic microflocs.

The stream then flows through a canister packed with stainless steel wire and a magnetic field is applied. The stainless steel wool captures the flocs by magnetic forces.

Ozonation

Ozone has been used continuously for nearly 80 years in municipal water treatment and the disinfection of water supplies. This practice began in France, then extended to Germany, Holland, Switzerland, and other European countries, and in recent years to Canada.

Ozone is a strong oxidizing substance with bactericidal properties similar to those of chlorine. In test conditions it was shown that the destruction of bacteria was between 600 and 3,000 times more rapid by ozone than by chlorine. Further, the bactericidal action of ozone is relatively unaffected by changes in pH while chlorine efficacy is strongly dependent on the pH of the water.

Ozone's high reactivity and instability as well as serious obstacles in producing concentrations in excess of 6 percent preclude central production and distribution with its associated economies of scale.

In the electric discharge (or corona) method of generating ozone, an alternating current is imposed across a discharge gap with voltages between 5 and 25 kV, and a portion of the oxygen is converted to ozone. A pair of large-area electrodes are separated by a dielectric (1–3 mm in thickness) and an air gap (approximately 3 mm). Although standard frequencies of 50 or 60 cycles are adequate, frequencies as high as 1,000 cycles are also employed. The mechanism for ozone generation is the excitation and acceleration of stray electrons within the high-voltage field. The alternating current causes the electron to be attracted first to one electrode and then to the other. As the electrons attain sufficient velocity, they become capable of splitting some oxygen molecules into free radical oxygen atoms. These atoms may then combine with O_2 molecules to form O_3.

Besides the disinfection of sewage effluent, ozone is used for sterilizing industrial containers such as plastic bottles, where heat treatment is inappropriate. Breweries use ozone as an antiseptic in destroying pathogenic ferments without affecting the yeast. It is also used in swimming pools and aquariums. It is sometimes used in the purification and washing of shellfish and in controlling slimes in cooling towers. Ozone has also been shown to be quite effective in destroying a variety of refractory organic compounds.

Ultraviolet Radiation

It has been shown that:

1. Ultraviolet radiation around 254 mm renders bacteria incapable of reproduction by photochemically altering the DNA of the cells.
2. A fairly low dose of ultraviolet light can kill 99 percent of the fecal coliform and fecal streptococcus.
3. Bacterial kill is independent of the intensity of the light but depends on the total dose.

4. Simultaneous treatment of water with UV and ozone results in higher microorganism kill than independent treatment with both W and ozone.

5. When ultrasonic treatment was applied before treating with the UV light, a higher bacteria kill was obtained.

6. The UV dose required to reduce the survival fraction of total coliform and fecal streptococcus to 102 (99 percent removal) is approximately 4 × 10 ff Einsteins/ml (see Figure 6.3).

7. Rough laboratory cost estimates indicate a cost of $0.002 to $0.004/m^3 for a UV dose of 4 × 10^{-8} Einsteins/ml.

Some limitations are associated with UV radiation for disinfection:

1. The process performance is highly dependent on the efficacy of upstream devices that remove suspended solids.

2. Another key factor is that the UV lamps must be kept clean in order to maintain their peak radiation output.

3. A further drawback is associated with the fact that a thin layer of water (<0.5 cm) must pass within 5 cm of the lamps.

One way of implementing the UV disinfection process at existing activated sludge plants involves suspending the UV lights (in the form of low-pressure mercury arc UV lamps with associated reflectors) above the secondary clarifiers. The effluent is exposed to the UV radiation as it rises over the wire in a thin film.

Electron Beam

The idea of using ionizing radiation to disinfect water is not new. Ionizing radiations can be produced by various radioactive sources (radioisotopes), by X-ray and particle emissions from accelerators, and by high-energy electrons. The advances in reliable, relatively low-cost devices for producing high-energy electrons are more significant.

Unlike X-rays and gamma rays, electrons are rapidly attenuated. The maximum range of a 1-million-volt electron is about 4 m in air and about 5 cm in water. In transit in matter, an electron loses energy through collisions that ionize atoms and molecules along its path. Bacteria and viruses are destroyed by the secondary ionization products produced by the primary traversing electron.

The energetic electrons dissociate water into free radicals H+ and OH. These may combine to form active molecules—hydrogen, peroxides, and ozone. These highly active fragments and molecules attack living structures to promote their oxidation, reduction, dissociation, and degradation.

Studies have indicated that 400,000 rads would be adequate for sewage disinfection. At 100 ergs per gram rad, 400,000 rads would raise the temperature of the water or sludge by 1°C. At this dose, each cm^2 of moving sludge would receive about 12×10^{12} electrons, each electron producing some 30,000 secondary ionizations.

Gamma Radiation

No operating gamma ray water disinfection system has been found in the United States. A number of experimental tests have been conducted.

Microwave

Microwaves have been used for dewatering in a number of applications (food concentration, high-quality paper drying). It is not surprising that attempts are made to use microwaves for sludge dewatering.

Laser

No application of laser technology has been found for disinfection, but it could conceivably be used for that purpose. When a laser beam impacts on biological material, a variety of physical and chemical reactions may take place depending on the wavelengths of the laser and the power density used. For wavelengths in the red and infrared region, the main effect is thermal, the absorbed energy being converted into heat. Laser illumination in the ultraviolet range may also photochemically excite molecules besides the thermal effect. Lasers can initiate photodynamic action in biological materials. This process is enhanced by the irradiation of dye sensitizers. These compounds may form long-life metastable complexes where the energy from the absorbed photon is available to catalyze biochemical reactions.

BIOLOGY OF AQUATIC SYSTEMS

Before examining the various techniques for purifying water, an understanding of the key biological organisms is necessary. These key organisms include bacteria, algae, protozoa, crustaceans, and fish. Bacteria and protozoa are the major groups of microorganisms. There are a number of waterborne diseases of man caused by bacteria. Some of these organisms are used in evaluating the sanitary quality of water for drinking and recreational purposes.

Waterborne Diseases

There are a number of infectious, enteric (that is, intestinal) diseases of man which are transmitted through fecal wastes. Pathogens (disease-producing agents) include bacteria, viruses, protozoa, and parasitic worms. Widespread diseases generally occur in regions where sanitary disposal of human feces is not practiced.

The most common waterborne bacterial diseases are typhoid fever (*Salmonella typhosa*), *Asiatic cholera* (vibrio comma), and bacillary dysentery (*Shigella dysenteriae*). The first of these is an acute infectious disease. Symptoms of typhoid fever are high fever and infection of the spleen, gastrointestinal tract, and blood. For cholera, symptoms include diarrhea, vomiting, and severe dehydration. Dysentery produces diarrhea, bloody stools, and high fever. These diseases can cause death and are still prevalent in many underdeveloped nations. However, in this country, proper environmental control has virtually eliminated these problems.

Waterborne outbreaks of infectious hepatitis have occurred. However, the main transmission mechanism is by person-to-person contact. The probability of outbreaks from municipally treated water supplies is low. Symptoms include loss of appetite, nausea, fatigue, and pain. Also, a yellowish color appears in the white of eyes and skin (yellow jaundice is an older term for the disease). It is generally not fatal except to individuals with weaker or older metabolisms.

Amoebic dysentery is the most common enteric protozoal infection. It is caused by Endamoeba histolytica and is transmitted by direct contact, food, and through the water in tropical climates. It is not transmittable via water in temperate climates. Disinfection, as with all these diseases, is the safest means of prevention.

Bilharziasis (or Schistosomiasis) is a parasitic disease generated by a small, flat worm that can infest the internal organs, such as the heart, lungs, and liver, and even the veins. Eggs of these worms existing in human abdominal organs can be transmitted to water via fecal discharges. Once in water, they hatch into miracida and enter into snails. They then develop into sporocysts that produce fork-tailed cercariae which eventually abandon their shells and attach onto humans. They bore through the skin, enter the bloodstream, and eventually find their way to the internal organs to establish their homes. There is no immunization for this disease. Many feel it is one of the world's worst health problems, particularly in agricultural regions of Africa and South America. Fortunately, this disease does not occur in the United States (the intermediate snail host just happens to be one of several specific species not found on the continental United States).

Nomenclature and Properties of Bacteria

Bacteria consists of simple, colorless, one-celled plants that utilize soluble food. They are capable of self-reproduction without the aid of sunlight. As decomposers, they represent decaying organic matter in nature. They typically range in size from $0.5–5\mu$ and as such are only visible through a microscope. Individual bacteria cells take on various geometries. Typical configurations include spheres, rods, or spirals. They may be single, in pairs, packets, or chains.

Reproduction is by binary fission, meaning a cell divides into two new cells, each of which matures and divides again. Fission takes place every 1,530 mill under ideal conditions. Ideal conditions mean that the growth environment has abundant food, oxygen, and essential nutrients.

Bacteria are named according to a binomial system. The first word is the genus and the second is the species name. The most frequently referred to bacterium in the sanitary field is Escherichia coli. E. coli is a common coliform that can be used as an indicator of water's bacteriological quality. Under a microscope and magnified 1,000 times, cells appear as individual short rods.

There are two major classifications of bacteria called heterotrophic and autotrophic. Heterotrophs, also called saprohytes, utilize organic substances both as a source of energy and carbon. Heterotrophs are further subclassified into three groups. Subclassifications are based on the bacteria's action toward free oxygen. Aerobes need free dissolved oxygen to decompose organics to derive energy for growth and reproduction. This can be described by Equation 2.1.

$$\text{Organics} + \text{Oxygen} \rightarrow CO_2 + H_2O + \text{Energy} \tag{2.1}$$

The second subgroup are anaerobes, which oxidize organics in the absence of dissolved oxygen. This is accomplished by using the oxygen which is found in other compounds (such as nitrate and sulfate). Anaerobic behavior can be described by the following reactions:

$$\text{Organics} + NO_3 \rightarrow CO_2 + N_2 + \text{Energy} \tag{2.2}$$

$$\text{Organics} + SO_4 = \rightarrow CO_2 + H_2S + \text{Energy} \tag{2.3}$$

Note also that:

$$\text{Organics} \rightarrow \text{Organic} \quad S + CO_2 + H_2O + \text{Energy} \tag{2.4A}$$

and that the organic acids undergo further reaction:

$$\text{Organic Acids} \rightarrow CH_4 + CO_2 + \text{Energy} \tag{2.4B}$$

Facultative bacteria comprise the last group and use free dissolved oxygen when available. However, they can also survive in its absence (that is, they also gain energy from the anaerobic reaction).

Heterotrophic bacteria decompose organics to obtain energy for the synthesis of new cells, respiration, and motility. Some energy is lost in the process as heat (see Figure 2.2).

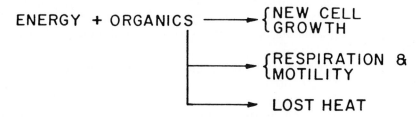

Figure 2.2 Products of cell synthesis.

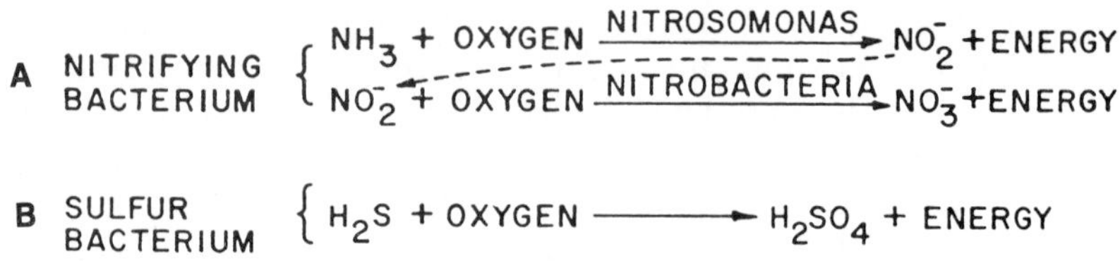

$$\text{A} \quad \begin{matrix} \text{NITRIFYING} \\ \text{BACTERIUM} \end{matrix} \left\{ \begin{matrix} NH_3 + \text{OXYGEN} \xrightarrow{\text{NITROSOMONAS}} NO_2^- + \text{ENERGY} \\ NO_2^- + \text{OXYGEN} \xrightarrow{\text{NITROBACTERIA}} NO_3^- + \text{ENERGY} \end{matrix} \right.$$

$$\text{B} \quad \begin{matrix} \text{SULFUR} \\ \text{BACTERIUM} \end{matrix} \left\{ H_2S + \text{OXYGEN} \longrightarrow H_2SO_4 + \text{ENERGY} \right.$$

Figure 2.3   Two forms of autotrophic bacteria.

Among the reactions described, Equation 2.1 describes the case where the greatest amount of energy is biologically available from a given quantity of matter. The lowest energy yield is from Equation 2.4, the strict anaerobic metabolism. Microorganisms found in wastewater strive for the maximum energy yield to maximize synthesis.

Autotrophic bacteria oxidize inorganic constituents for energy and utilize carbon dioxide as a source of carbon. The major bacteria types in this class are nitrifying, sulfur, and iron bacteria. Nitrifying bacteria will oxidize ammonium nitrogen to nitrate (see Figure 2.3A).

Sulfur bacteria perform the reaction given in Figure 2.3B. This reaction causes crown corrosion in sewers. Water in sewers quite frequently turns septic and generates hydrogen sulfide gas, as in Equation 2.3. The H_2S generated absorbs in the condensation moisture on the sewer side walls and the crown of the pipe. Those sulfur bacteria able to survive at very low pH (pH < 1) oxidize weak H_2S acid to strong sulfuric acid. This oxidation reaction depletes the oxygen from the sewer air. Crown corrosion of concrete-lined systems can greatly reduce the structural integrity of piping and eventually cause walls to collapse.

Iron bacteria oxidize soluble inorganic ferrous iron to insoluble ferric (Figure 2.4). Certain types of filamentous bacteria (Leptothrix and Crenothrix) deposit oxidized iron in the form of $Fe(OH)_3$ in their sheath. This produces yellow or reddish-colored slimes. Water pipes are ideal environments for these type bacteria as they have an abundance of highly dissolved iron content to provide energy and bicarbonates to serve as a carbon source. As these microorganisms mature and die, they decompose, generating obnoxious odors and foul tastes.

Fungi and Molds

Fungi are microscopic nonphotosynthetic plants which include in their classification yeast and molds. Yeasts have a commercial value as they are used for fermentation operations in distilling and brewing. When anaerobic con-

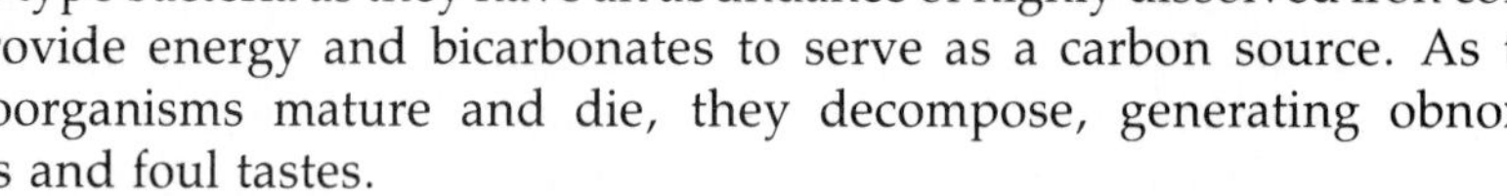

$$\begin{matrix} \text{IRON} \\ \text{BACTERIUM} \end{matrix} \left\{ Fe^{++}(\text{FERROUS}) + \text{OXYGEN} \longrightarrow Fe^{+++}(\text{FERRIC}) + \text{ENERGY} \right.$$

Figure 2.4   Iron bacteria oxidize soluble inorganic ferrous iron to insoluble ferric.

ditions exist, yeasts metabolize sugar, manufacturing alcohol from the synthesis of new cells. Alcohol is not manufactured under aerobic conditions and the yield of new yeast cells is greater.

Filamentous forms of fungi are molds. These best resemble higher orders of plant life, having branched or threadlike growths. They grow best in environments consisting of acid solutions with high sugar concentrations. Molds are nonphotosynthetic, multicellular, heterotrophic, and aerobic. The growth of molds can be suppressed by increasing the pH.

Algae, Protozoa, and Multicellular Animals

Algae are microscopic photosynthetic plants. They are among the simplest plant forms, having neither roots, stems, nor leaves. Algae typically range from single-cell entities (which impart a green color to surface waters) to branched forms that can be seen by the naked eye. The latter often appear as attached green slime on surface bodies of water. *Diatoms* refers to single-celled algae which are housed in silica shells. The blue-green algae generally associated with water pollution are Anacystis, Anabaena, and Aphanizomellon. Green algae are Oocystis and Pediastrum.

Algae are autotrophic; that is, they use carbon dioxide or bicarbonates as sources of carbon. Inorganic nutrients of phosphate and nitrogen as ammonia or nitrate are also used. Some trace nutrients are also necessary (magnesium, boron, cobalt, calcium).

The reaction or process by which algae propagate is known as photosynthesis.

$$CO_2 + PO_4 + NH_3 \xrightarrow[\text{Dark Reaction}]{\text{Solar Energy}} \text{New Cell Growth} + O_2 \qquad (2.5)$$

The products of photosynthesis are new plant growth and oxygen. The energy supplied to the reaction is derived from sunlight. Pigments biochemically convert solar energy into useful energy for plant reproduction and survival. In prolonged absence of sunlight, plant matter performs a dark reaction to exist. In this case, algae absorb oxygen and degrade stored food to produce yield energy for respiratory functions. The reaction rate for the dark reaction is much slower than photosynthesis.

Macrophytes are aquatic photosynthetic plants (excluding algae). They often appear on surface bodies of water as floating, submerged, and immersed aggregates. Floating plants are not anchored or rooted.

In the animal kingdom, one of the simplest forms is the protozoan. Protozoa are single-celled aquatic animals that have relatively complex digestive systems. They use solid organic material as food and multiply by binary fission. They are aerobic organisms and digest bacteria and algae and, consequently, play an essential role in the aquatic food chain. The smallest type are the flagellated protozoa which range in size from 10μ to 50μ. These have long hairlike strands which provide motility by a whiplike action. The amoeba is a member of the protozoa family.

Rotifiers are simple, multicelled, aerobic animals. These metabolize solid food. Rotifiers are found in natural waters, stabilization ponds, and extended aeration basins in municipal treatment plants.

Crustaceans are multicellular animals (about 2 mm in size). They are herbivores which ingest algae and are in turn eaten by fish.

Biological Growth Factors

Major factors affecting biological growth are temperature, nutrient availability, oxygen supply, pH, degree of sunlight, and the presence of toxins. Bacteria are classified by their optimum temperature range for growth. For example, mesophilic bacteria grow best in an optimum temperature range for growth. For example, mesophilic bacteria grow best in a temperature range of 10–40°C (optimum at 37°C). In general, the rate of biological activity almost doubles for every 10–15°C rise in temperature within the range of 5–35°C. This is illustrated in Figure 2.5. Beyond 40°C, mesophilic activity drops dramatically and thermophilic growth is initiated (thermophilic bacteria have a range between 45–75°C, with an optimum of about 55°C). Thermophilic bacteria are typically more sensitive to temperature variations.

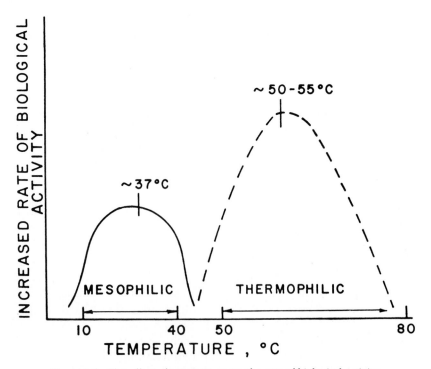

Figure 2.5   The effect of temperature on the rate of biological activity.

Water Quality Test Methods

Determination of the bacteriological quality of water is not a straightforward analysis. The testing for a specific pathogenic bacteria can often lead to erroneous conclusions. Analyses for pathogenic bacteria are difficult to perform. In general, data are not quantitatively reproducible. As an example, if Salmonella was found to be absent from a water sample, this does not exclude the possible presence of Shigella, Vibrio, or disease-producing viruses. The bacteriological quality of water is based on test procedures for nonpathogenic indicator organisms (principally the coliform group).

Coliform bacteria, typified by Escherichia coli and fecal streptococci (enterococci), reside in the intestinal tract of man. These are excreted in large numbers in the feces of humans and other warm-blooded animals. Typical concentrations average about 50,000,000 coliforms per gram. Untreated domestic wastewater generally contains more than 3,000,000 coliforms per 100 ml. Pathogenic bacteria and viruses causing enteric diseases originate from the same source (that is, fecal discharges of diseased persons). Consequently, water contaminated by fecal pollution is identified as being potentially dangerous by the presence of coliform bacteria.

Standards for drinking water specify that a water is safe provided that the test method does not reveal more than an average of one coliform organism per 100 ml. The number of pathogenic bacteria, such as Salmonella typhosa, in domestic wastewater is generally less than 1 per mil coliforms, and the average density of enteric viruses has been measured as a virus-to-coliform ratio of 1:100,000. The die-off rate of pathogenic bacteria is greater than the death rate of coliforms outside of the intestinal tract of animals. Consequently, upon exposure to treatment, a reduction in the number of pathogens relative to coliforms will occur. Water quality based on a standard of less than one coliform per 100 ml is statistically safe for human consumption. That is, there is a high improbability of ingesting any pathogens. This is an Environmental Protection Agency (EPA) standard applicable only to processed water where treatment includes chlorination.

Coliform criteria for body-contact water use and recreational use have been established by most states. Upper limits of 200 fecal coliforms per 100 ml and 2,000 total coliforms per 100 ml have been established. These values are only guidelines since there is no positive epidemiological evidence that bathing beaches with higher coliform counts are associated with transmission of enteric diseases. Some experts feel that these standards may be too conservative from a standpoint of realistic public health risk. In recent years, no cases of enteric disease have been linked directly to recreational water use in this country. Coliform standards applied to water used for swimming are linked to water-associated diseases of the skin and respiratory passages rather than enteric diseases. This, naturally, is entirely different than the purpose of the coliform standard for drinking water, which is related to enteric disease transmission. Here tighter restrictions are imperative, since a water distribution system has the potential of mass transmission of pathogens in epidemic proportions.

Water sample collection techniques differ depending on the source being tested. The minimum number of water samples collected from a distribution system which are examined each month for coliforms is a function of the population. For example, the minimum number required for populations of 1,000 and 100,000 are 2 and 100, respectively. To ascertain compliance with the bacteriological requirements of drinking water standards, a certain number of positive tests must not be exceeded. When 10-ml standard portions are examined, not more than 10 percent in any month should be positive (that is, the upper limit of coliform density is an average of one per 100 ml).

Coliforms are defined as all aerobic and facultative anaerobic, nonspore-forming species. Gram-stain negative rods ferment lactose and produce gas within 48 h of incubation (at 35°C). The initial coliform analysis is the presumptive test which is based on gas production from lactose. In this test, 10-ml portions of water samples are transferred into prepared fermentation tubes using sterile pipettes. The tubes contain lactose or lauryl tryptose, broth, and

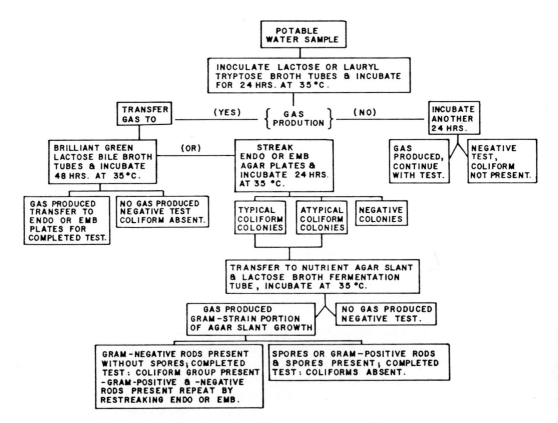

Figure 2.6 Outline of tests for coliform detection in potable water systems.

inverted vials. Inoculated tubes are placed in a warm-air incubator (at 35°C ± 0.5°C). Growth with the production of gas (the gas is identified by the presence of bubbles in the inverted vial) means a positive test. That is, it indicates that coliform bacteria may be present. A negative reaction, either no growth or growth without gas, excludes coliforms.

Such tests are employed to substantiate or refute the presence of coliforms in a positive presumptive test. The two procedures normally used in testing are outlined in Figure 2.6. In normal, potable water coliform testing, the test is confirmed using brilliant green bile broth. Occasionally, one may desire to run a completed test, as outlined in Figure 2.6. This involves transferring a colony from an Endo (or EMB plate) to nutrient agar and into lactose broth. If gas is not produced in the lactose fermentation tube, the colony transferred did not contain coliforms and the test is negative. If gas is generated, a portion of growth on the nutrient agar is smeared onto a glass slide and prepared for observation under a microscope using the Gram-stain technique. If the bacteria are short rods, with no spores present, and the Gram-stain is negative, the coliform group is present and the test is completed. If the culture Gram-stains positive (purple color), the completed test is negative.

In examining surface water quality, an elevated-temperature coliform test is used to separate microorganisms of the coliform group into those of fecal and nonfecal sources. This approach is applicable to studies of stream pollution, raw-water sources, wastewater treatment systems, bathing waters, and general water quality monitoring. It is not recommended as a substitute for the coliform tests used in examination of potable waters. Details of the test procedure are outlined in Figure 2.7.

The water analysis is incomplete unless the number of coliform bacteria present is determined as well. A multiple-tube fermentation technique can be used to enumerate positive presumptive, confirmed, and fecal coliform tests. Results of the tests are expressed in terms of the most probable number (MPN). That is, the count is based on a statistical analysis of sets of tubes in a series of serial dilutions. MPN is related to a sample volume of 100 ml. Thus, an MPN of 10 means 10 coliforms per 100 ml of water.

For MPN determination, sterile pipettes calibrated in 0.1-ml increments are used. Other equipment includes sterile screw-top dilution bottles containing 99 ml of water and a rack containing six sets of five lactose broth fermentation tubes. A sterile pipette is used to transfer 1.0-ml portions of the sample into each of five fermentation tubes. This is followed by dispensing 0.1 ml into a second set of five. For the next higher dilution (the third), only 0.01 ml of sample water is required. This small quantity is very difficult to pipette accurately, so 1.0 ml of sample is placed in a dilution bottle containing 99 ml of sterile water and mixed. The 1.0-ml portions containing 0.01 ml of the surface water sample are then pipetted into the third set of five tubes. The fourth set receives 0.1 ml from this same dilution bottle. The process is then carried one more step by transferring 1.0 ml from the first dilution bottle into 99 ml of water in the second for another hundredfold dilution. Portions

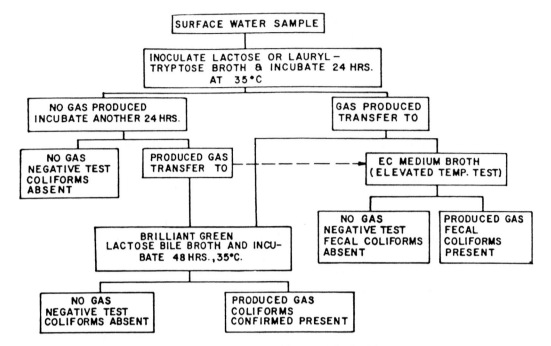

Figure 2.7 Outline of tests for coliform and fecal-coliform groups.

from this dilution bottle are pipetted into the fifth and sixth tube sets. After incubation (48 h at 35°C), the tubes are examined for gas production and the number of positive reactions for each of the serial dilutions is recorded. Figure 2.8 describes the technique.

A final testing technique worth noting is the membrane filter method for coliform testing. This procedure involves passing a measured water sample through a membrane filter to remove the bacteria. The filter is then placed on a growth medium in a petri dish. The bacteria retained by the filter pad grow and establish a small colony. The number of coliforms present is established by counting the number of colonies and expressing this value in terms of number per 100 ml of water. This technique has been widely adopted for use in water quality monitoring studies, especially since it requires considerably less laboratory apparatus than the standard multiple-tubes technique. Also, this technique can be adapted to field studies.

Equipment needed to perform the membrane filter coliform test includes filtration units, filter membranes, absorbent pads, forceps, and culture dishes. The common laboratory filtration unit consists of a funnel that fastens to a receptacle bearing a porous plate to support the filter membrane. The filter-holding assembly can be constructed of glass, porcelain, or stainless steel. It is sterilized by boiling, autoclaving, or ultraviolet radiation. For filtration, the assembly is mounted on a side-arm filtering flask which is evacuated to draw the sample through the filter. For field use, a small hand-sized plunger pump

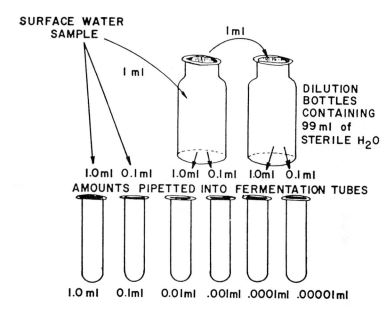

Figure 2.8   Procedure for preparation of the multitube fermentation method for evaluating MPN of coliform bacteria in surface waters.

or syringe is used to draw a sample of water through the small assembly holding the filter membrane.

Commercial filter membranes are normally 2-in diameter disks with pore openings of 0.45 ($\pm$0.02) μ. This is small enough to retain microbial cells. Filters used in determining bacterial counts have a grid printed on the surface. To facilitate counting colonies, the filter membranes must be sterilized prior to use, either in a glass petri dish or wrapped in heavy paper. After sterilization, the pads are placed in culture dishes to absorb the nutrient media on which the membrane filter is placed. During the testing, filters are handled on the outer edges with forceps that are also sterilized before use.

Glass or disposable plastic culture dishes are used. If glass petri dishes are employed, a humid environment must be maintained during incubation. This prevents losses of media by evaporation (the dishes have loose-fitting covers). Disposable plastic dishes have tight-fitting lids which minimize the problem of dehydration.

The size of the filtered sample is established by the anticipated bacterial density. An ideal quantity results in the growth of about 50 coliform colonies and not more than 200 colonies of all types. Often it may be difficult to anticipate the number of bacteria in a sample. Two or three volumes of the same sample must be tested. When the portion being filtered is less than 20 ml, a small amount of sterile dilution water is added to the funnel before filtration. This uniformly disperses the bacterial suspension over the entire surface of the filter. The filter-holding assembly is placed on a suction flask. A sterile filter is placed grid side up over the porous plate of the apparatus using sterile forceps. The funnel is then locked in place holding the membrane. Filtration is performed by passing the sample through the filter under partial vacuum. A culture dish is prepared by placing a sterile absorbent pad in the upper half of the dish and pipetting enough enrichment media on top to saturate the pad. M-Endo medium is used for the coliform group and M-FC for fecal coliforms. The filter is then removed from the filtration apparatus and placed directly on the pad in the dish. The cover is replaced and the culture is incubated (for 24 h at 35°C). For fecal coliforms, incubation is performed by placing the culture dishes in watertight plastic bags and submerging them in a water bath at 44.5°C. Coliform density is calculated in terms of coliforms per 100 ml by multiplying the colonies counted by 100 and dividing this value by the milliliters of the sample filtered.

DISINFECTION BY CHLORINATION

Disinfection has received increased attention over the past several years from regulatory agencies through the establishment and enforcement of rigid bacteriological effluent standards. In upgrading existing wastewater treatment facilities, the need for improved disinfection as well as the elimination of odor problems are frequently encountered. Adequate and reliable disinfection is essential in ensuring that wastewater treatment plants are both environmentally safe and aesthetically acceptable to the public.

Chlorine is the most widely used disinfectant in water and wastewater treatment. It is used to destroy pathogens, control nuisance microorganisms, and for oxidation. As an oxidant, chlorine is used in iron and manganese removal, for destruction of taste and odor compounds, and in the elimination of ammonia nitrogen. It is, however, a highly toxic substance and recently concerns have been raised over handling practices and possible residual effects of chlorination. Recent shortages and price escalation of liquid chlorine have also emphasized the need to consider alternative methods of disinfection.

Properties of Chlorine

Chlorine (Cl_2) is a greenish-yellow-colored gas having a specific gravity of 2.48 as compared to air under standard conditions of temperature and pressure. It was discovered in 1774 from the chemical reaction of manganese dioxide ($MnNO_2$) and hydrochloric acid (HCl) by the Swedish chemist, Scheele, who believed it to be a compound containing oxygen. In 1810, it was named by Sir Humphrey Davy, who insisted it was an element (from the Greek work *chloros*, meaning greenish-yellow). In nature, it is found in the combined state only, usually with sodium as salt ($NaCl$), carnallite ($KMgCl_36H_2O$), and sylvite (KCl).

Chlorine is a member of the halogen (salt-forming) group of elements and is derived from chlorides by the action of oxidizing agents and, most frequently, by electrolysis. As a gas, it combines directly with nearly all elements. At 10°C, 1 volume of water dissolves about 3.10 volumes of chlorine; at 30°C, only 1.77 volumes of Cl_2 are dissolved in 1 volume of water.

In addition to being the most widely used disinfectant for water treatment, chlorine is extensively used in a variety of products, including paper products, dyestuffs, textiles, petroleum products, pharmaceuticals, antiseptics, insecticides, foodstuffs, solvents, paints, and other consumer products. Most chlorine produced is used in the manufacture of chlorinated compounds for sanitation, pulp bleaching, disinfectants, and textile processing. It is also used in the manufacture of chlorates, chloroform, and carbon tetrachloride and in the extraction of bromine. Among other past uses, chlorine served as a war gas during World War I.

As a liquid, chlorine is amber colored and is 1.44 times heavier than water. In solid form, it exists as rhombic crystals. Various properties of chlorine are given in Table 2.3.

Chlorine gas is a highly toxic substance, capable of causing death or permanent injury due to prolonged exposures via inhalation. It is extremely irritating to the mucous membranes of the eyes and the respiratory tract. It will combine with moisture to liberate nascent oxygen to form hydrochloric acid. If both these substances are present in quantity, they can cause inflammation of the tissues with which they come in contact. Pulmonary edema may result if lung tissues are attacked.

Chlorine gas has an odor detectable at a concentration as low as 3.55 ppm. Irritation of the throat occurs at 15 ppm. A concentration of 50 ppm is

TABLE 2.3   PROPERTIES OF CHLORINE

Symbol	Cl
(as gas)	Cl_2
Atomic Number	17
Atomic Weight	35.453
Melting Point (°C)	$-101.$
Boiling Point (°C)	-34.5
Liquid Density (0°C and 3.65 atm; g/l)	1.47
Vapor Pressure (mmHg @ 20°C)	4800
Vapor Density (@ STP: g/l)	2.49
Viscosity (micropoises) at	
T = 12.7°C	129.7
= 20°C	132.7
= 50°C	146.9
= 100°C	167.9
= 150°C	187.5
= 200°C	208.5

TABLE 2.4 ACUTE TOXICITY DATA

	Concentration (ppm)	Exposure Time (hr)
Inhalation TC_{LO} (humans)	15	—
Inhalation LD_{LO} (humans)	430	0.5
Inhalation LC_{50} (rats)	293	1

LC_{50} = lethal concentration to 50 percent of a specified population.
LD_{LO} = lowest published lethal dose.
TC_{LO} = lowest published toxic concentration.

considered dangerous for even short exposures. At or above concentrations of 1,000 ppm, exposure may be fatal. Table 2.4 gives acute toxicity data.

Chlorine can also cause fires or explosions upon contact with various materials. Table 2.5 lists various substances chlorine can react with to create fire hazards. It emits highly toxic fumes when heated and reacts with water or steam to generate toxic and corrosive hydrogen chloride fumes.

Chlorine Chemistry

In the United States, chlorine was first used as a disinfectant for municipal wastewater treatment in the Jersey City, New Jersey, Boonton reservoir in 1908. This also marked the first legal recognition of chlorine as a disinfectant for public health protection.

Chlorine is a strong oxidizing agent and can be used to modify the chemical character of water. For example, it is used to control bacteria, algae, and macroscopic biological-fouling organisms in condenser cooling towers. It is also used to alter the chemical character of some industrial process waters, such as the destruction of sulfur dioxide and ammonia, the reduction of iron and manganese, and the reduction of color (examples include bleaching operations in the pulp and paper industry and oxidation of organic constituents).

In water chlorine hydrolyzes to form hypochlorous acid (HOCl), as shown by the following reactions:

$$Cl_2 + H_2O = HOCl + H^+Cl^- \qquad (2.6)$$

The hypochlorous acid undergoes further ionization to form hypochlorite ions (OCl^-):

$$HOCl = H^+ + OCl^- \qquad (2.7)$$

Equilibrium concentrations of HOCl and OCl depend on the pH of the wastewater. Increasing the pH shifts the preceding equilibrium relationships to the right, causing the formation of higher concentrations of HOCl.

Chlorine may also be applied as calcium hypochlorite and sodium hypochlorite. Hypochlorites are salts of hypochlotous acid. Calcium hypochlorite ($Ca(OCl)_2$) represents the predominant dry form used in the United

TABLE 2.5 MATERIALS THAT REACT WITH CHLORINE TO CAUSE FIRES/EXPLOSIONS

Turpentine	Al	CS	$(H_2O + KOH)$
Ether	Sb	$CSHC_2$	I_2
Ammonia Gas	As	Co_2O	Hydroxylamine
Illuminating Gas	AsS_2	Cs_3N	Fe
Hydrocarbons	AsH_3	$(C + Cr(OCl)_2)$	FeC_2
Hydrogen and Powdered Metals	Ba_3P_2	Cu	Li
Polydimethylsiloxane	Bi	CuC_2	Li_2C_2
Polypropylene	B	Dialkyl Phosphines	Li_6C_2
Drawing Wax	BPI_2	Diborane	Mg
Rubber	B_2S_3	Dibutyl Phathalate	Mn
Sulfamic Acid	Brass	$Zn(C_2H_5)_2$	Mn_3P_2
$As_2(CH_3)_4$	BrF_5	C_2H_6	HgO
UC_2	Ca	C_2H_4	HgS
Acetaldehyde	$(CaC_2 + KOH)$	Ethylene Imine	Hg
C_2H_2	$Ca(ClO_2)_2$	$C_2H_5PH_2$	Hg_3P_2
Alcohols	Ca_3N_2	F_2	CH_4
Akyl Isothiourea	Ca_3P_2	Ge	Nb
Alkyl Phosphines	Cs_2	Glycerol	NI_3
	OF_2	KH	SbH_3
	H_2SiO	Ru	Sr_3P
	$(OF_2 + Cu)$	$RuHC_2$	Te
	PH_3	Si	Th
	P	SiH_2	Sn
	$P(SNC)_3$	Ag_2O	WO_2
	P_2O_3	Na	U
	PCB's	$NaHC_2$	V
	K	Na_2C_2	Zn
	KHC_2	SnF_2	ZrC_2

States. Calcium hypochlorite is commercially available in granular powdered or tablet forms. Either of these forms readily dissolves in water and contains approximately 70 percent available chlorine.

Sodium hypochlorite (NaOCl) is commercially available in liquid form at concentrations typically between 5 percent to 15 percent available chlorine. Hypochlorites react in water as follows:

$$NaOCl \rightarrow Na^+ OCl^- \qquad (2.8)$$

$$Ca(OCl)_2 \rightarrow Ca^{+2} + 2OCl^- \qquad (2.9)$$

$$H^+ OCl^- \, HOCl \qquad (2.10)$$

The amount of HOCl plus OCl in wastewater is referred to as the free available chlorine.

Chlorine is a very active oxidizing agent and is therefore highly reactive with readily oxidized compounds such as ammonia. Chlorine readily reacts with ammonia in water to form chloramines.

$$HOCl + NH_3 \rightarrow H_2O + NH_2Cl \text{ (monochloramine)} \qquad (2.11)$$

$$HOCl + NH_2Cl \rightarrow H_2O + NHCl_2 \text{ (dichloramine)} \qquad (2.12)$$

$$HOCl + NHCl_2 \rightarrow H_2O + NCl_3 \text{ (trichloramine)} \qquad (2.13)$$

The specific reaction products formed depend on the pH of the water, temperature, time, and the initial chlorine-to-ammonia concentration ratio. In general, monochloramine and dichloramine are generated in the pH range of 4.5 to 8.5. Above pH 8.5, monochloramine usually exists alone. However, below pH 4.4, trichloramine is produced.

When chlorine is mixed with water containing ammonia, the residuals developed produce a curve similar to the one shown in Figure 2.9. The positive sloped line from the origin represents the concentration of chlorine applied or the residual chlorine if all of that applied appears as residual. The solid curve represents chlorine residuals corresponding to various dosages that remain after some specified contact time. The chlorine demand at a specified dosage is obtained from the vertical distance between the applied and residual curves. Chlorine demand represents the amount of chlorine reduced in chemical reactions (that is, it is the amount that is no longer available).

For molar chlorine to ammonia-nitrogen ratios below 1, monochloramine and dichloramine are formed with their relative amounts dependent on pH and other factors. When higher dosages of chlorine are added, the chlorine-

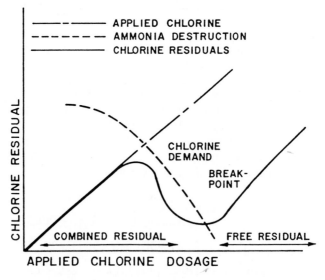

Figure 2.9 Characteristic breakpoint chlorination curve.

to-nitrogen ratio increases, resulting in an oxidation of the ammonia and a reduction of the chlorine. Specifically, three moles of chlorine will react with two moles of ammonia, generating nitrogen gas and reducing chlorine to the chloride ion:

$$2NH_3 + 3Cl_2 \rightarrow N_2 + 6HCl \qquad (2.14)$$

Residuals of chloramine decline to a minimum value that is referred to as the breakpoint. When dosages exceed the breakpoint, free chloride residuals result. Breakpoint curves are unique for different water samples since the chlorine demand is a function of the concentration of ammonia, presence of other reducing agents, and the contact time between chlorine application and residual testing.

Germicidal Destruction

Chlorine's ability to destroy bacteria and various microorganisms results from chemical interference in the functioning of the organism. Specifically, it is the chemical reaction between HOCl and the bacterial or viral cell structure which inactivates the required life processes. The high germicidal efficiency of HOCl is attributed to the ease by which it is able to penetrate cell walls. This penetration is comparable to that of water and is due both to its low molecular weight (that is, it's a small molecule) and its electrical neutrality.

Organism fatalities result from a chemical reaction of HOCl with an enzyme system in the cell which is essential to the metabolic functioning of the organism. The enzyme attacked is triosephosphate dehydrogenase, found in most cells and essential for digesting glucose. Other enzymes also undergo attack. However, triosephosphate dehydrogenase is particularly sensitive to oxidizing agents.

The OCl$^-$ ion resulting from the dissociation is a relatively poor disinfectant because of its inability to diffuse through a microorganism's cell walls. This is because of its negative charge.

The sensitivity of bacteria to chlorination is well known. However, the effect on protozoans and viruses has not been entirely delineated. Protozoal cysts and enteric viruses are more resistant to chlorine than are coliforms and other enteric bacteria. However, very little evidence exists to indicate that current water treatment practices are inadequate (no outbreaks of viral or protozoal infections have been reported and waterborne diseases attributed to these pathogens are rare in this country).

Contact Time, pH, and Temperature Effects

Hypochlorous acid and hypochlorite ion are known as free available chlorine. The chloramines are known as combined available chlorine and are slower than free chlorine in killing microorganisms. For identical conditions of contact time, temperature, and pH in the range of 6 to 8, it takes at least 25 times more combined available chlorine to produce the same germicidal efficiency. The difference in potency between chloramines and HOCl can be explained by the difference in their oxidation potentials, assuming the action of chloramine is of an electrochemical nature rather than one of diffusion, as seems to be the case for HOCl.

The effect of pH alone on chlorine efficiency is shown in Figure 2.10. Chlorine exists predominantly as HOCl at low pH levels. Between pH of 6.0 and 8.5, a dramatic change from undissociated to completely dissociated hypochlorous acid occurs. Above pH 7.5, hypochlorite ions prevail; while above 9.5, chlorine exists almost entirely as OCl. Increased pH also diminishes the disinfecting efficiency of monochloramine.

It has also been demonstrated that the germicidal effectiveness of free and combined chlorine is markedly diminished with decreasing water temperature. In any situation in which the effects of lowered temperature and high pH value are combined, reduced efficiency of free chlorine and chloramines is marked.

These factors directly affect the exposure time needed to achieve satisfactory disinfection. Under the most ideal conditions, the contact time needed with free available chlorine may only be on the order of a few minutes; combined available chlorine under the same conditions might require hours.

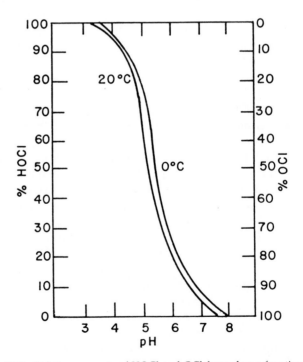

Figure 2.10   Relative amounts of HOCl and OCl formed as a function of pH.

TABLE 2.6   RECOMMENDED CHLORINE DOSAGE RANGES

Wastewater Type	Chlorine Dosage (mg/l)
Raw Sewage	6–12
(Septic) Raw Sewage	12–25
Settled Sewage	5–10
Chemical Precipitation Effluent	3–10
Trickling Filter Effluent	3–10
Activated Sludge Effluent	2–8
Sand Filter Effluent	1–5

Chlorine Dosage Rates and Residuals

Table 2.6 gives recommended ranges of chlorine dosages for disinfection of various wastewaters. Recommended minimum bactericidal chlorine residuals are given in Table 2.7. Data in Table 2.7 are based on water temperatures between 20°C to 25°C after a 10-minute contact for free chlorine and a 60-minute contact for combined available chlorine. The minimum residuals required for cyst destruction and inactivation of viruses are much greater. Although chlorine residuals in Table 2.7 are generally adequate, surface waters from polluted waterways are usually treated with much heavier chlorine dosages.

Ordinary chlorination will destroy all strains of coli, aerogenes, pyocyaneae, typhsa, and dysenteria. In addition to these microorganisms, three other types are readily destroyed:

- Enteric vegetative bacteria (Eberthella, Shigella, Salmonella and Vibrio species).
- Worms such as the blook flukes (Schistosoma species).
- Viruses (for example, the virus of infectious hepatitis).

Each of these groups of organisms differs in its reaction with chlorine. There is evidence that the comparative reaction of different organisms to one form of chlorine is not necessarily maintained relative to other forms.

TABLE 2.7   MINIMUM BACTERICIDAL CHLORINE RESIDUALS (MG/L)

pH Value	Free Available Chlorine Residual After 10-min Contact	Combined Available Chlorine Residual After 60-min Contact
6.0	0.2	1.0
7.0	0.2	1.5
8.0	0.4	1.8
9.0	0.8	>3.0

Chlorination Systems

Water chlorination is carried out by using both free and combined residuals. The latter involves chlorine application to produce chloramine with natural or added ammonia. Anhydrous ammonia is used if insufficient natural ammonia is present in the wastewater. Although the combined residual is less effective than free chlorine as a disinfectant, its most common application is as a posttreatment following free residual chlorination to provide initial disinfection.

Free residual chlorination establishes a free residual through the destruction of naturally present ammonia. High dosages of chlorine applied during treatment may result in residuals that are esthetically objectionable or undesirable for industrial water uses. Dechlorination is sometimes performed to reduce the chlorine residual by adding a reducing agent (called a dechlor). Sulfur dioxide is often used as the dechlor in municipal plants. Aeration by submerged or spray aerators also diminishes the residual chlorine concentration.

The chlorine used for disinfection is available in three forms: liquified compressed gas, calcium hypochlorite or sodium hypochlorite, and chlorine bleach solutions. Liquid chlorine is shipped in pressurized steel cylinders with sizes typically 100 and 500 lb; one-ton containers are used in large installations. There are two types of chlorine dispensing systems: direct feed and solution feed. The first involves metering dry chlorine gas and conducting it under pressure to the water. Solution-feed systems meter chlorine gas under vacuum and dissolve it in a small amount of water, forming a concentrated solution which is then applied to the water being treated. At 20°C, 1 volume of water dissolves 2.3 volumes of chlorine gas (about 7,000 mg/l). At concentrations of total chlorine below 1,000 mg/l, none of the gas exists in solutions as Cl_2; all of it is present as HOCl or dissociated ions.

Calcium hypochlorite is a dry bleach which is available in granular and tablet forms. Calcium hypochlorite is relatively stable under normal conditions; however, it can undergo reactions with organic materials. It should be stored in an isolated area. Sodium hypochlorite is available in liquid form. It is marketed in carboys and rubber-lined drums for small quantities. Sodium hypochlorite solutions are highly corrosive, unstable, and require storage at temperatures below 85°F. Sodium hypochlorite can either be delivered to the site in liquid form in 500–5,000-gallon tank cars or trucks, or manufactured on site. It is normally sold at a concentration of 12 percent to 15 percent by weight of available chlorine. It can be manufactured on site from salt or from sea water.

The main component in a chlorine gas feed is the variable orifice inserted in the feed line to control the rate of flow out of the cylinder. The orifice basically consists of a grooved plug sliding in a fitted ring. Feed rate is adjusted by varying the V-shaped opening. Since a chlorine cylinder pressure varies with temperature, the discharge through such a throttling valve does not remain constant without frequent adjustments of the valve setting. Also, conditions on the outlet side vary with pressure changes at the point of application. Therefore, a pressure-regulating valve is used between the cylinder and the orifice, with a vacuum-compensating valve on the discharge side. A safety pressure-relief valve is held closed by vacuum.

Chlorine feeders can be controlled either manually or automatically based on flow or chlorine residual, or both. In manual mode, a continuous feed rate is established. This is satisfactory when chlorine demand and flow are relatively constant and where operators are available to make adjustments. Automatic proportional control equipment is used to adjust the feed rate to provide a constant preestablished dosage for all rates of flow. This is accomplished by metering the main flow and using a transmitter to signal a chlorine feeder. An analyzer located downstream from the point of application is used to monitor the chlorinator. Combined automatic flow and residual control maintain a present chlorine residual in the water that is independent of the demand and flow variations. The feeder is designed to respond to signals from both the flow meter transmitter and the chlorine residual analyzer.

For hypochlorite solutions, positive-displacement diaphragm pumps (either mechanically or hydraulically actuated) are used. The hypochlorinator consists of a water-powered pump paced by a positive-displacement water meter. The meter register shaft rotates proportionately to the main line flow and controls a cam-operated pilot valve. This in turn regulates water now discharged of hypochlorite that is proportional to the main flow. Admitting main pressure behind the pumping diaphragm balances the water pressure in the pumping head. The advantage of this system is that the pump does not need electrical power. The hypochlorite dosage can be manually adjusted by changing the stroke length setting of the pump.

Chlorine Contact Tanks

The configuration of contact tanks can result in appreciable differences between actual and theoretical contact times. Contact times and germicidal efficiency depend on a number of parameters, the most important being the mixing characteristics of the basin. Proper designs must account for possible flow pattern elimination via short circuiting, acceptable dosage rates, optimum pH range, and upstream removal of ammonia nitrogen.

Rapid dispersement of chlorine at the addition point increases chlorine contact and improves disinfection efficiency. Baffles can be designed to generate turbulence at the chlorine addition point and improve mixing. Baffled systems have the advantage of not requiring mechanical equipment. Mechanical mixing or air agitation can be used where plant hydraulics will not allow the use of baffles, or where a portion of the existing basin can be converted to a mixing chamber and the remainder of the basin and/or a long outfall sewer can be used to provide the needed contact time.

Toxic Effects of Chlorine

The toxicity of chlorine residuals to aquatic life has been well documented. Studies indicate that at chlorine concentrations in excess of 0.01 mg/l, serious hazard to marine and estuarine life exists. This has led to the dechlorination of wastewaters before they are discharged into surface water bodies. In addition to being toxic to aquatic life, residuals of chlorine can produce halogenated organic compounds that are potentially toxic to man. Trihalomethanes (chloroform and bromoform), which are carcinogens, are produced by chlorination.

Chlorine Dioxide

Chlorine dioxide, discovered in 1811 by Davy, was prepared from the reaction of potassium chlorate with hydrochloric acid. Early experimentation showed that chlorine dioxide exhibited strong oxidizing and bleaching properties. In the 1930s, the Mathieson Alkali Works developed the first commercial process for preparing chlorine dioxide from sodium chlorate. By 1939, sodium chlorite was established as a commercial product for the generation of chlorine dioxide.

Chlorine dioxide uses expanded rapidly in the industrial sector. In 1944, chlorine dioxide was first applied for taste and odor control at a water treatment plant in Niagara Falls, New York. Other water plants recognized the uses and benefits of chlorine dioxide. In 1958, a national survey determined that 56 U.S. water utilities were using chlorine dioxide. The number of plants using chlorine dioxide has grown more slowly since that time.

At present, chlorine dioxide is primarily used as a bleaching chemical in the pulp and paper industry. It is also used in large amounts by the textile industry, as well as for the bleaching of flour, fats, oils, and waxes. In treating drinking water, chlorine dioxide is used in this country for taste and odor control, decolorization, disinfection, provision of residual disinfectant in water distribution systems, and oxidation of iron, manganese, and organics. The principal use of chlorine dioxide in the United States is for the removal of taste and odor caused by phenolic compounds in raw water supplies.

Chlorine dioxide is a yellow-green gas and soluble in water at room temperature to about 2.9 g/l chlorine dioxide (at 30 mm mercury partial pressure) or more than 10 g/l in chilled water. The boiling point of liquid chlorine dioxide is 11°C; the melting point is -59°C. Chlorine dioxide gas has a specific gravity of 2.4. The oxidant is used in a water solution and is five times more soluble in water than chlorine gas. In addition, chlorine dioxide does not react with water in the same manner that chlorine does. Chlorine dioxide is volatile; consequently, it can be stripped easily from a water solution by aeration.

Chlorine dioxide has a disagreeable odor, similar to that of chlorine gas, and is detectable at 17 ppm. It is distinctly irritating to the respiratory tract at a concentration of 45 ppm in air. Concentrations above 11 percent can be mildly explosive in air. As a gas or liquid, it readily decomposes upon exposure to ultraviolet light. It is also sensitive to temperature and pressure, two reasons why chlorine dioxide is generally not shipped in bulk concentrated quantities.

Chlorine dioxide has a much greater oxidative capacity than chlorine and is therefore a more effective oxidant in lower concentrations. Chlorine dioxide also maintains an active residual in potable water longer than chlorine does. It does not react with ammonia or with trihalomethane precursors when prepared with no free residual chlorine.

Chlorine dioxide is prepared from feedstock chemicals by several methods. The specific method depends on the quantity needed and the safety limitations in handling the various feedstock chemicals. The most common processes are:

FROM SODIUM CHLORITE ($NaClO_2$):

- Acid and sodium chlorite.
- Gaseous chlorine and sodium chlorite.
- Sodium hypochlorite, acid, and sodium chlorite.

FROM SODIUM CHLORATE ($NaClO_3$):

- The sulfur dioxide process.
- The methanol process.

The first group of processes is more commonly used. The second group of processes is frequently used by industry where the quantities produced are much greater than in water utilities.

Oxidation of phenols with chlorine dioxide or chlorine produces chlorinated aromatic intermediates before ring rupture. Oxidation of phenols with either chlorine dioxide or ozone produces oxidized aromatic compounds as intermediates which undergo ring rupture upon treatment with more oxidant and/or longer reaction times. In many cases, the same nonchlorinated, ring-ruptured aliphatic products are produced using ozone or chlorine dioxide.

In oxidizing organic materials, chlorine dioxide can revert back to the chlorite ion. In the presence of excess chlorine (or other strong oxidant), chlorite can be preoxidized to chlorine dioxide. Using large excesses of chlorine dioxide over the organic materials appears to favor oxidation reactions (without chlorination), but slight excesses appear to favor chlorination.

When excess free chlorine is present with the chlorine dioxide, chlorinated organics usually are produced, but in lower yields, depending on the concentration of chlorine and its reactivity with the particular organic(s) involved. Treatment of organic compounds with pure chlorine dioxide containing no excess free chlorine produces oxidation products containing no chlorine in some cases, but products containing chlorine in others.

Under drinking water plant treatment conditions, humic materials and/ or resorcinol do not produce trihalomethanes with chlorine dioxide even when a slight excess of chlorine (1 percent to 2 percent) is present. Also, saturated aliphatic compounds are not reactive with chlorine dioxide. Alcohols are oxidized to the corresponding acids.

The gaseous chlorine-sodium chlorite process for producing chlorine dioxide uses aqueous chlorine and aqueous sodium chlorite to produce a mixture of chlorine dioxide and chlorine (commonly as HOCl). Figure 2.11 shows such a system, consisting of a chlorine dioxide generator, a gas chlorinator, a storage reservoir for liquid sodium chlorite, and a chemical metering pump. (Sodium chlorite solution can be prepared from commercially available dry chemical by adding it to water.) The recommended feed ratio of chlorine to sodium chlorite is 1:1 by weight. Additional chlorine can be injected into the reactor vessel without changing the overall production of chlorine dioxide.

A major disadvantage of this system is the limitation of the single-pass gas-chlorination phase. Unless increased pressure is used, this equipment is unable to achieve higher concentrations of chlorine as an aid to a more complete and controllable reaction with the chlorite ion. The French have developed a variation of this process using a multiple-pass enrichment loop on the chlorinator to achieve a much higher concentration of chlorine and thereby quickly attain the optimum pH for maximum conversion to chlorine dioxide. This system is illustrated in Figure 2.12. By using a multiple-pass recirculation system, the chlorine solution concentrates to a level of 5–6 g/l. At this concentration, the pH of the solution reduces to 3.0 and thereby provides the low pH level necessary for efficient chlorine dioxide production. A single pass results in a chlorine concentration in water of about 1 g/l, which produces a

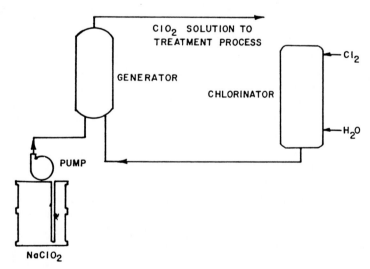

Figure 2.11 Illustrates the major components of a gaseous sodium chlorite-chlorine dioxide generation system.

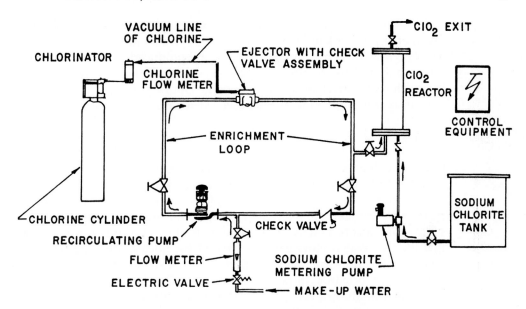

Figure 2.12   French system for generating chlorine dioxide.

pH of 4 to 5. If sodium chlorite solution is added at this pH, only about 60 percent yield of chlorine dioxide is achieved. The remainder is unreacted chlorine (in solution) and chlorite ion. When upwards of 100 percent yield of chlorine dioxide is achieved, there is virtually no free chlorite or free chlorine carrying over into the product water.

The French system can be designed for variable-feed rates with automatic control by an analytical monitor. This has the advantages of eliminating the chlorine dioxide storage reservoir. Production can be varied by 20 equal increments. A 10 kg/h (530 lb/day) reactor can be varied in 0.5 kg/h (26.5 lb/day) steps over the range of 0–10 kg/h, and this can be accomplished by automatic control with the monitor located in the main plant control panel.

Another approach to chlorine dioxide production is the acid-sodium chlorite system. The combination of acid and sodium chlorite produces an aqueous solution of chlorine dioxide without production of significant amounts of free chlorine. The acid-based process avoids the problem of differentiating between chlorine and chlorine dioxide for establishing an oxidant residual. This system uses liquid chemicals as the feedstock. Each tank has a level sensor to avoid overfilling. The tanks are installed below ground in concrete bunkers which are capable of withstanding an explosion. There are no floor drains in these bunkers. Any spillage must be pumped with corrosion-resistant pumps. Primary and backup sensors with alarms warn of any spillage. Because of the potential explosiveness, chemicals are diluted prior to the production of chlorine dioxide. The dilution is carried out on a batch basis controlled by level monitors. Proportionate quantities of softened dilution water along with the chemical reagents are pumped to mixing vessels by means of calibrated double-metering pumps. After the reactor is properly filled, an agitator within the container mixes the solution. Dilutions of 9 percent HCl and 7.5 percent sodium chlorite are produced in the chemical preparation process. The chlorine dioxide is subsequently manufactured on a batch basis. The final strength of the solution is about 20 percent, 90 percent to 95 percent of this is chlorine dioxide and 4 percent to 7 percent is chlorine.

Summary

Chlorine is the most widely used disinfectant in water treatment. It appears, however, that it may not be the best disinfectant to use for drinking water where poor-quality raw water or completely recycled water is used. Other reasons for considering alternative disinfection techniques include the possibility that disinfection by chloramines will allow viruses to remain viable or that the inactivated virus particles have viable nucleic acids that may be released within humans, the reduction of germicidal efficiency with elevated pH, and the formation of persistent chlorinated organic compounds.

Chlorine dioxide has proven to be a strong oxidizing agent. When free of chlorine, it does not form trihalomethane compounds in drinking water. It is less likely than chlorine to form chlorinated compounds with most organics commonly encountered in raw water supplies. Chlorine dioxide is effective in oxidizing organic complexes of iron and manganese, imparts no taste and odor to treated water, and provides a highly stable, long-lasting oxidant residual.

DISINFECTION WITH INTERHALOGENS AND HALOGEN MIXTURES

The interhalogen compounds are the bromine- and iodine-base materials. It is the larger, more positive halogen that is the reactive portion of the interhalogen molecule during the disinfection process. Although only used on a limited basis at present, there are members of this class that show great promise as environmentally safe disinfectants.

Properties of Bromine and Bromides

Bromine (from the Greek word *bromos*, meaning stench) has an atomic weight 79.909, atomic number 35, melting point $-7.2°C$, and boiling point 58.78°C. As a gas it has a density of 7.59 g/l and as a liquid 3.12 g/l (20°C). The element was discovered by Balard in 1826 but not prepared in quantity until 1860. It is a member of the halogen group of elements. Bromine is found mainly in the bromide form, widely distributed and in relatively small proportions. Extractable bromides occur in the ocean and salt lakes, brines, or saline deposits left after these waters evaporated during earlier geological periods. The average bromide content of ocean water is 65 ppm by weight (about 308,000 tons of bromine per cubic mile of sea water). The Dead Sea is one of the richest commercial sources of bromine in the world (containing nearly 0.4 percent at the surface and up to 0.6 percent at deeper levels). In the United States, major sources of bromine are the brine wells in Arkansas, Ohio, and Michigan (bromide contents range from 0.2 percent to 0.4 percent).

Bromine is the only liquid nonmetallic element. It is a heavy, mobile, reddish-brown liquid that readily volatilizes at room temperature to a red vapor having a strong pungent odor. Its disagreeable odor strongly resembles chlorine and has a very irritating effect on the eyes and throat. Bromine is readily soluble in water or carbon disulfide, forming a red solution. It is less active than chlorine but more so than iodine. Bromine unites readily with many elements and has a bleaching action.

The toxic action of bromine is similar to that of chlorine and can cause physiological damage to humans through inhalation and oral routes. It is an irritant to the mucous membranes of the eyes and upper respiratory tract. Severe exposures may result in pulmonary edema. Chronic exposure is similar to therapeutic ingestion of excessive bromides.

It is considered a moderate fire hazard. As liquid or vapor, it can enter spontaneous chemical reactions with reducing materials. It is a very powerful oxidizer. Table 2.8 lists a number of compounds bromine is known to react violently with. Bromine is considered a highly dangerous material. Upon being heated, it emits highly toxic fumes. It will react with water or steam to produce toxic and corrosive fumes.

The most common inorganic bromides are sodium, potassium, ammonium, calcium, and magnesium bromides. Methyl and ethyl bromides are among the most common organic bromides. The inorganic bromides produce a number of toxic effects in humans: depression, emaciation, and in severe

TABLE 2.8   REDUCING MATERIALS THAT REACT WITH BROMINE

Acetaldehyde	Cs_2O	Dimethyl formamide
C_2H_2	Ss_2C_2	Ethyl phosphine
Acrylonitrile	CsC_2H	Isobutyrophenone
Aluminum	ClF_3C_2	Li
NH_3	CuH_2	Li_2C_2
Lead	Cu_2C_2	Li_2Si_2
Boron	F_2	Mg_3P_2
Nitrogen	Ge	CH_3OH
$Ni(CO)_4$	Olefins	O_3
NI_3	OF_2	PH_3
K	PO_X	P
Rb_2C_2	RbC_2H	AgN_3
NaC_2H	Na_2C_2	Na
Sr_3P	Sn	UC_2
ZrC_2		

cases, psychoses and mental deterioration. Bromide rashes (called bromo-derma) can occur especially on the facial area and resemble acne and furun-culosis. This often occurs when bromide inhalation or administration is pro-longed. Organic bromides such as methyl bromide and ethyl bromide are volatile liquids of relatively high toxicity. When any of the bromides are strongly heated, they emit highly toxic fumes.

Interhalogen Compounds and Their Properties

Interhalogen compounds are formed from two different halogens. These com-pounds resemble the halogens themselves in both their physical and chemical properties. Principal differences show up in their electronegativities. This is clearly shown by the polar compound ICl, which has a boiling point almost 40°C above that of bromine, although both have the same molecular weights.

Interhalogens have bond energies that are lower than halogens and therefore in most cases they are more reactive. These properties impart special germicidal characteristics to these compounds. The principal germicidal com-pound of this group is bromine chloride.

At equilibrium, BrCl is a fuming dark red liquid below 5°C. It exists as a solid only at relatively low temperatures. Liquid BrCl can be vaporized and metered as a vapor in equipment similar to that used for chlorine.

BrCl is prepared by the addition of equivalent amounts of chlorine to bromine until the solution has increased in weight by 44.3 percent: Equation 2.15 shows the reaction.

$$Br_2 + Cl_2 \quad\quad 2BrCl \tag{2.15}$$

BrCl can be prepared by the reaction in the gas phase or in aqueous hydrochloric acid solution. In the laboratory, BrCl is prepared by oxidizing bromide salt in a solution containing hydrochloric acid.

$$kBrO_3 + 2kBr + 6HCl \rightarrow 3BrCl + 3kCl + 3H_2O \tag{2.16}$$

BrCl exists in equilibrium with bromine and chlorine in both gas and liquid phases as described by Equation 2.15. Table 2.9 lists various physical constants of BrCl.

Due to the polarity of BrCl, it shows greater solubility than bromine in polar solvents. In water, it has a solubility of 8.5 gms per 100 gms of water at 20°C (that is, 2.5 times the solubility of bromine; 11 times that of chlorine). Bromine chloride's solubility in water is increased greatly by adding chloride ions to form the complex chlorobromate ion, $BrCl_2$.

Chemistry of Bromine Chloride

Various organic and inorganic species that act as reducing agents react with and destroy free halogen residuals during interaction with microorganisms (see Figure 2.13 for examples of competitive reactions).

TABLE 2.9   PHYSICAL CONSTANTS OF BrCl

Molecular weight	115.37
Melting point (°C)	−66
Boiling point (°C)	5
Density (g/cc), 20°C	2.34
Heat of fusion (cal/g)	17.6
Heat of vaporization (cal/g)	53.2
Heat formation (kcal/mole)	0.233
Heat capacity (cal./deg. mole, 298°k)	8.38
Entropy (cal./deg. mole, 298°k)	57.34
Dipole moment	0.56
Electrical conductivity ($dm^{-1}\ cm^{-1}$)	—
Degree of dissociation (%, vapor 25°C)	21

Competitive reactions depend on the reactivity of the chemical species, temperature, contact time, and pH. The quality of the effluent and the method of adding the disinfectant also help determine the specific reaction pathways.

Bromine chloride is about 40 percent dissociated into bromine and chlorine in most solvents. Because of its high reactivity and fast equilibrium, BrCl often generates products that result almost entirely from it. This is illustrated by the disinfectant products shown in Figure 2.14. The major portion of the BrCl is eventually reduced to inorganic bromides and chlorides, with the exception of addition and substitution reactions with organic constituents.

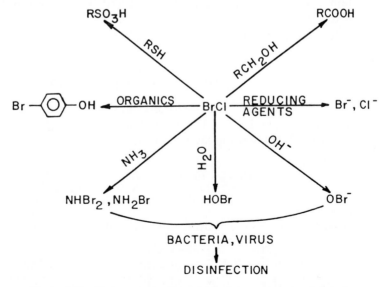

Figure 2.13   Various competitive reactions in wastewater disinfection.

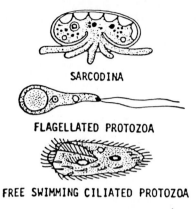

SARCODINA

FLAGELLATED PROTOZOA

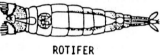

FREE SWIMMING CILIATED PROTOZOA

STALKED CILIATED PROTOZOA

ROTIFER

Figure 2.14 Typical microorganisms important in the biological treatment process.

It should be noted that although BrCl is mainly a brominating agent that is competitive with bromine, its chemical reactivity makes its action similar to that of chlorine (that is, disinfection, oxidation, and a bleaching agent).

BrCl hydrolyzed exclusively to hypobromous acid, according to Equation 2.17; and if any hydrobromic acid (HBr) is formed by hydrolysis of the dissociated bromine, it quickly oxidizes to hydrobromous acid via hypochlorous acid (Equation 2.18).

$$BrCl + H_2O + HOBr + HCl \qquad (2.17)$$

$$HBr + HOCl \rightarrow HOBr + HCl \qquad (2.18)$$

The hydrolysis constant for BrCl in water is compared to that of bromine and chlorine in Table 2.10. As shown, the hydrolysis constant for BrCl in water is 42,000 times greater than the hydrolysis of bromine in water.

TABLE 2.10 COMPARISON OF HYDROLYSIS CONSTANTS OF BrCl TO Br AND Cl IN WATER AT 0°C

	Hydrolysis Constant	No. Times BrCl Greater Than
BrCl	2.94×10^{-5}	1.0
Br	0.70×10^{-9}	42,000
Cl	1.45×10^{-4}	0.203

Since hypohalous acid is a much more active disinfectant than the hypohalite ion, the effect of pH on ionization becomes important. Hypobromous acid has a lower ionization value than hypochlorous acid and this contributes to the higher disinfectant activity of BrCl compared with chlorine.

Bromine chloride also undergoes very specific reactions with ammonia and with organics. Monobromamine and dibromamine are the major products formed from reactions between BrCl and ammonia. These are unstable compounds in most conventional wastewater treatment plant effluents. In comparing the activities of bromamine versus chloramines, the effects of ammonia and high pH tend to improve the bromamine performance whereas the chloramine activity is reduced significantly. The reaction of ammonia with either BrCl or chlorine to form the halamine is very fast and generally goes to completion. As such, the presence of ammonia is essential to the disinfectant properties. Most sewage effluents typically have high ammonia concentrations in the range of 5–20 ppm. For such samples, the predominant bromine species (pH at 7 to 8) monobromamine and dibromamine are approximately equally distributed.

There are a large number of organics that undergo disinfection during the purification process. There are unfortunately a number of undesirable by-products and side reactions which occur with some of them. One is the reaction between chlorine and phenol, producing chlorophenols, which are suspected carcinogens. Chlorophenols have obnoxious tastes and are toxic to aquatic life even at very low concentrations. Brominated phenolic products which are formed in the chlorobromination of wastewater are generally more readily degraded and often less offensive than their chlorinated counterparts.

The major organic reactions of BrCl consist of electrophilic brominations of aromatic compounds. Many aromatic compounds do not react in aqueous solution unless the reaction involves activated aromatic compounds (an example being phenol). Bromine chloride undergoes free-radical reactions more readily than bromine.

Metal ions in their reduced state also undergo reactions with BrCl. Examples include iron and manganese.

$$Fe^{+2} + BrCl \rightarrow Fe^{+3} + Br^- + Cl^- \tag{2.19}$$

$$Mn^+ BrCl \rightarrow Mn^{+2} + Br^- + Cl^- \tag{2.20}$$

Wastewater occasionally contains hydrogen sulfide and nitrites. These contribute to higher halogen demands. Many of these reactions reduce halogens to halide salts.

Bromine chloride's reactivity with metals is not as great as that of bromine; however, it is comparable to chlorine. Dry BrCl is typically two orders of magnitude less reactive with metals than dry bromine. Most BrCl is less corrosive than bromine. Like chlorine, BrCl is stored and shipped in steel containers. Also, Kynar® and Viton plastics and Teflon® are preferred over polyvinyl chloride (PVC) when BrCl is in the liquid or vapor states.

Disinfection with Bromine Chlorine

In chlorination, chlorine's reaction with ammonia forms chloramines, greatly reducing its bactericidal and virucidal effectiveness. The biocidal activity of monochloramine is only 0.02–0.01 times as great as that of free chlorine. Typical ammonia concentrations found in secondary sewage range from 5–20 ppm, which is about an order of magnitude greater than the amount needed to form monochloramine from normal chlorination dosages (which requires about 5–10 ppm). Therefore, monochloramine is the major active chlorine constituent in chlorinated sewage plant effluents. In contrast, BrCl ammonia reactions produce the major product bromamines. Bromamines have disinfectant characteristics which are significantly different than chloramines.

Toxicity of Aquatic Life

Bromamines are considerably less stable than chloramines in receiving waters. Bromamines tend to break down into relatively harmless constituents typically in under 60 minutes. Consequently, BrCl is less damaging to marine life than chlorine.

Chloramines at concentrations below 0.1 ppm have resulted in fish kills. There are also indirect effects from chloramine contamination. For example, fish populations tend to avoid toxic regions, even at very low levels of concentrations. Consequently, large areas of receiving waters can become unavailable to many species of fish and even cause blockage of upstream migrations during the spawning season.

It should be noted that although chlorine efficiency is increased by nitrification, BrCl performance is not. Because of the high biocidal activity of bromamines, it is not necessary to utilize high concentrations and breakpoint conditions to achieve active halogen residuals, as is the case in chlorination. The breakpoint reaction with BrCl is achieved almost immediately in the presence of even slight excess amounts of bromine at pH levels of 7 to 8. There is, however, no need to reach the breakpoint to achieve good disinfectant properties with BrCl. In contrast, with chlorine it is necessary to add amounts in excess of the breakpoint to obtain sterilizing characteristics.

PROPERTIES OF IODINE

Iodine (from the Greek, *iodines*, meaning violet) has an atomic weight of 126.9044, atomic number 53, melting point 113.5°C, and boiling point 184.35°C. As a gas, its density is 11.27 g/l and as a solid its specific gravity is 4.93 (20°C). This halogen was discovered by Courtois in 1811.

It occurs sparingly in the form of iodides in sea water from which it is assimilated by seaweeds, in Chilean saltpeter and nitrate-bearing soil, in

TABLE 2.11 COMPOUNDS WITH
WHICH IODINE WILL REACT VIOLENTLY

C_2H_2	$CsHC_2$	F_2	P	ZrC
Al	Cs_2C_2	Li	k	
NH_3	Cs_2O	Li_2C_2	$RbHC_2$	
NH_4OH	Cl_2	Li_6C	Rb_2C_2	
Sb	ClF_3	Mg	AgN_3	
BrF_5	Cu_2C_2	OF_2	NaH	

brines from ancient sea deposits, and in brackish waters derived from oil and salt wells. Pure grades of iodine can be obtained from the reaction of potassium iodide with copper sulfate.

Iodine is a grayish-black, lustrous solid that volatilizes at ordinary temperatures to a blue-violet gas. It forms compounds with many elements. However, it is less active than many of the other halogens which displace it from iodides. Iodine dissolves readily in chloroform, carbon tetrachloride, and carbon disulfide. It is only slightly soluble in water.

Iodine is highly irritating to the skin, eyes, and mucous membranes. Its effect on the human body is similar to that of bromine and chlorine. However, it is more irritating to the lungs. Table 2.11 gives a partial list of compounds with which iodine will react.

Disinfection with Iodine Compounds

Two interhalogens having strong disinfecting properties are iodine monochloride (ICl) and iodine bromide (IBr). Iodine monochloride has found use as a topical antiseptic. It may be complexed with nonionic or anionic detergents to yield bactericides and fungicides that can be used in cleansing or sanitizing formulations. These generally have a polymer structure which establishes its great stability, increased solubility, and lower volatility. By reducing the free halogen concentration in solution, polymers reduce both the chemical and bactericidal activity. Complexes of ICl are useful disinfectants which compromise lower bactericidal activity with increased stability.

Iodine monochloride is itself a highly reactive compound, reacting with many metals to produce metal chlorides. Under normal conditions it will not react with tantalum, chromium, molybdenum, zirconium, tungsten, or platinum. With organic compounds, reactions cause iodination, chlorination, decomposition, or the generation of halogen addition compounds. In water, ICl hydrolyzes to hypoiodous and hydrochloric acids. In the absence of excess chloride ions, hypoiodous acid will disproportionate into iodic acid and iodine.

$$ICl + H_2O \; HOI + HCl \tag{2.21}$$

$$5HOI \rightarrow HIO_3 + 2I_2 + 2H_2O \tag{2.22}$$

Iodine bromide has a chemistry similar to ICl. Iodine bromide reacts with aromatic compounds to produce iodination in polar solvents and bromination in nonpolar solvents. It has complex chemical properties, as its solubility is increased more effectively by bromide than by chlorided ions. Primary hydrolysis takes place in the presence of hydrobromic acid:

$$IB_r + H_2O \leftrightharpoons HOI + HBr \tag{2.23}$$

In the absence of acid, the following occurs:

$$5IBr + 3H_2O \leftrightharpoons 2I_2 + HIO_3^- + 5HBr \tag{2.24}$$

As a disinfectant, IBr is used in its complexed or stabilized forms. Unfortunately, it undergoes hydrolysis and dissociation reactions in aqueous solutions, both reactions being major limitations. Its disinfecting properties are similar to ICl and as in the case of ICl, germicidal activity should not be reduced by haloamine formation since bromamines are highly reactive and iodoamines are not generated. Upon application of prepared solutions to control microorganisms, the complex releases IBr gradually. This process forms free iodine during the decomposition of IBr (the decomposition takes place as fast as the IBr is released).

Disinfection with Halogen Mixtures

Two approaches that have been investigated recently for disinfection are mixtures of bromine and chlorine, and mixtures containing bromide or iodide salts. Some evidence exists that mixtures of bromine and chlorine have superior germicidal properties than either halogen alone. It is believed that the increased bacterial activity of these mixtures can be attributed to the attacks by bromine on sites other than those affected by chlorine.

The oxidation of bromide or iodide salts can be used to prepare interhalogen compounds or the hypollalous acid in accordance with the following reaction:

$$HOCl + NaBr \rightarrow HOBr + HCl \tag{2.25}$$

It has been reported that the rate of bacterial sterilization by chlorine in the presence of ammonia is accelerated with small amounts of bromides. As little as 0.25 ppm of bromamines can be significant under some conditions. However, if chloramines are produced prior to contact with bromide ions, the reaction and subsequent effect are reduced. Improved germicidal activity has also been shown for mixtures containing bromides and iodides with various chlorine releasing compounds. Bromide improves the disinfecting properties of dichloroisocyanuric acid and hypochlorite against several bacteria. Bromine-containing compounds are useful for their combined bleaching and disinfectant properties.

Summary

There has been the concern that the use of interhalogen compounds in wastewater disinfection could produce unknown organic and inorganic halogen-containing substances. In the case of iodine, concern has been expressed over the physiological aspects in water supplies. Extensive studies have been reported on the role played by iodine and iodides in the thyroid glands of animals and man. Information on acute inhibition of hormone formation by excessive amounts of iodine is well known. Despite the fact that no strong evidence exists that iodine is harmful as a water disinfectant, only limited use has been attempted.

Chronic bromide intoxication from continuous exposure to dosages above 3–5 g is called bromism. Typical symptoms are skin rash, glandular excretions, gastrointestinal disturbances, and neurological disturbances. Bromide can be absorbed from the intestinal tract and contaminate the body in a manner very similar to that for chloride.

Brominated drinking water does not, however, significantly increase the amount of bromine admitted internally. The amount of additional bromine in chlorobrominated waters will not significantly increase human bromine concentrations nor result in bromism.

Figure 2.14 shows typical microorganisms that can be affected or destroyed by chemical disinfection.

OZONE TREATMENT

Ozone (O_3) is a powerful oxidant, and application to effluent treatment has developed slowly because of relatively high capital and energy costs compared to chlorine. Energy requirements for ozone are in the range of 10 to 13 kWh/lb generated from air, 4 kWh/lb from oxygen, and 5.5 kWh/lb from oxygen-recycling systems. Operating costs for air systems are essentially the electric power costs; for oxygen systems the cost of oxygen (2 to 3¢/lb) must be added to the electrical cost. Capital costs of large integrated ozone systems are $300 to $400 a pound per day of ozone generated and $100 a pound per day of ozone for the generator alone.

Actual uses of ozone include odor control, industrial chemicals synthesis, industrial water and wastewater treatment, and drinking water. Lesser applications appear in fields of combustion and propulsion, foods and pharmaceuticals, flue gas-sulfur removal, and mineral and metal refining. Potential markets include pulp and paper bleaching, power plant cooling water, and municipal wastewater treatment.

The odor control market is the largest and much of this market is in sewage treatment plants. Use of ozone for odor control is comparatively simple and efficient. The application is for preservation of environmental quality;

in addition, alternative treatment schemes requiring either liquid chemical oxidants (like permanganate or hydrogen perioxide) or incineration can significantly increase capital and costs.

Ozone applications in the United States for drinking water are few. However, the potential market is large, if environmental or health needs ever conclude that an alternate disinfectant to chlorine should be required. Although energy costs of ozonation are higher than those for chlorination, they may be comparable to combined costs of chlorination ·dechlorination-reaeration, which is a more equivalent technique. One of ozone's greatest potential uses is for municipal wastewater disinfection.

Technical, economic, and environmental advantages exist for ozone bleaching of pulp in the paper industry as an alternate to hypochlorite or chlorine bleaching which yields deleterious compounds to the environment.

Principles of Ozone Effluent Treatment

Ozone was first discovered by the Dutch philosopher Van Marun in 1785. In 1840, Schonbein reported and named ozone from the Greek word *ozein*, meaning to smell. The earliest use of ozone as a germicide occurred in France in 1886, when de Meritens demonstrated that diluted ozonized air could sterilize polluted water. In 1893, the first drinking water treatment plant to use ozone was constructed in Oudshorrn, Holland. Other plants quickly followed at Wiesbaden (1901) and Paderborn (1902) in Germany. In 1906, a plant in Nice, France, was constructed using ozone for disinfection. Today, there are over 1,000 drinking water treatment plants in Europe utilizing ozone for one or more purposes. In the United States, the first ozonation plant was constructed in Whiting, Indiana, in 1941 for taste and odor control.

Over 100 years ago it had been demonstrated that ozone (O_3), the unstable triatomic allotrope of oxygen, could destroy molds and bacteria and by 1892 several experimental ozone plants were in operation in Europe. In the 1920s, however, as a result of wartime research, during World War I, chlorine became readily available and inexpensive, and began to displace ozone as a purifier in municipalities throughout the United States. Most ozone studies and development were dropped at this time, leaving ozonation techniques, equipment, and research at a primitive stage. Ozone technology stagnated, and the development and acceptance of ozone for water and wastewater treatment was discontinued.

In addition to the popular use of chlorination as a wastewater disinfectant and the consequent technology lag in ozonation research, there was a third impediment to ozone commercialization: the comparatively high cost of ozonation in relation to chlorination. Ozone's instability requires on-site generation for each application, rather than centralized generation and distribution. This results in higher capital requirements, aggravated by a comparatively large electrical energy requirement. Ozone's low solubility in water and the generation of low concentrations, even under ideal conditions, also necessitates more elaborate and expensive contacting and recycling systems than chlorination.

In spite of such obstacles there is interest from time to time in the use of ozone, particularly for wastewater treatment. The technology for the destruction of organics and inorganics in water has not kept pace with the increasingly more sophisticated water pollution problems arising from greater loads, new products, and new sources of pollutant entry into the environment and increased regulation. The growing trend toward water reuse and the fact that some highly toxic pollutants may be refractory to conventional treatment methods has spurred investigation into new treatments, including ozonation.

A significant impetus from time to time for developing new methods is dissatisfaction with chlorination. Chlorine affects taste and odor and produces chloramines and a wide variety of other potentially hazardous chlorinated compounds in wastewaters. It seriously threatens the environment with an estimated 1,000 tons per year of chlorinated organic compounds discharged into U.S. waters (chloramines are not easily degradable and pose a hazard to the environment) and is questionable as a drinking water viricidal disinfectant. Ozone's development, on the other hand, could parallel a greater environmental awareness and a resulting demand for higher-quality effluents, as its potential for overcoming these problems is possible.

Properties of Ozone

Ozone (molecular weight is 48) is an unstable gas, having a boiling point of $-112°C$ (atmospheric pressure). Ozone is partially soluble in water (approximately 20 times more soluble than oxygen), and has a characteristic penetrating odor which is readily detectable at concentrations as low as 0.01–0.05 ppm. Ozone is the most powerful oxidant currently available for use for wastewater treatment. Commercial generation equipment generates ozone at concentrations of 1 percent to 3 percent in air (that is, 2 percent to 6 percent in oxygen). Ozone is unstable in water; however, it is more stable in air, especially in cool, dry air.

As a strong oxidant, ozone reacts with a wide variety of organics. Ozone oxidizes phenol to oxalic and acetic acids. It oxidizes trihalomethane (THM) compounds to a limited extent within proper pH ranges and reduces their concentration by air stripping. Trihalomethanes are also oxidized by ozone in the presence of ultraviolet light. Oxidation by ozone does not result in the formation of THMs as does chlorination.

A combination of ozone and ultraviolet light destroys DDT, malathion, and other pesticides. However, high dosages and extended contact times that are not normally encountered in drinking water treatment are needed. Ozonized organic substances are usually more biodegradable and absorbable than the starting, unoxidized substances. When ozonation is employed as the final treatment step for potable water systems in water containing significant concentrations of dissolved organics, bacterial regrowth in the distribution system can occur. Consequently, ozonation is not typically used as the final treatment step but rather followed by granular activated carbon filtration and sometimes by the addition of a residual disinfectant.

Humic materials are the precursors of THMs. Humic substances can be oxidized by ozonation. Under proper conditions significant reduction in THM formation can be realized when ozone is applied prior to a chlorination step.

Because of ozone's instability, it is able to produce a series of almost instantaneous reactions when in contact with oxidizable compounds. One example follows.

$$O_3 + 2KI + H_2O \rightarrow I_2 + O_2 + 2KOH \qquad (2.26)$$

In reaction (2.26), iodine is liberated from a solution of potassium iodide. This reaction can be used to assess the amount of ozone in either air or water. For determination in air or oxygen, a measured volume of gas is drawn through a wash bottle containing potassium iodide solution. Upon lowering the pH with acid, titration is effected with sodium thiosulfate, using a starch solution as an indicator. There is a similar procedure for determining ozone in water.

A typical ozone treatment plant consists of three basic subsystems: feed-gas preparation; ozone generation; and ozone/water contacting. Commercially, ozone is generated by producing a high-voltage corona discharge in a purified oxygen-containing feedgas. The ozone is then contacted with the water or wastewater; the treated effluent is discharged and the feedgas is recycled or discharged.

Ozone's high reactivity and instability, as well as serious obstacles in producing concentrations in excess of 6 percent, preclude central production and distribution with its associated economies of scale. The requirement for on-site generation and application of ozone must yield a cost-efficient, low-maintenance operation in order to be useful. The feedgas employed in ozonation systems is either air, oxygen, or oxygen-enhanced air. The particular selection of feedgas for each application is based on economics and depends on several factors: total quantity of ozone required; desired concentration of ozone in the feedgas; and fate (recycle or discharge) of the feedgas. For a given ozone generator with a specified power input and gas flow, two to three times as much ozone may be generated from oxygen as from air. The maximum concentration economically produced from air is about 2 percent, while that generated from pure oxygen is approximately 6 percent.

The use of higher concentrations of ozone provides two advantages: capital and operating costs per pound of ozone produced are substantially reduced, and a greater concentration gradient for mass transfer of ozone is provided in the contacting step, yielding increased ozone-utilization efficiency. These advantages, however, must be weighed against the increased cost of oxygen production. Air is generally employed in those applications requiring less than 50 pounds/day of low-concentration ozone. If air is the feedgas, it must be dried and cooled to reduce accumulation of corrosive nitric acid and nitrogen oxides that occur as by-products when the dew point is above 40°C.

Ozone may be produced by electrical discharge in an oxygen-containing feedgas or by photochemical action using ultraviolet light. For large-scale applications, only the electric-discharge method is practical since the use of ultraviolet energy produces only low-volume, low-concentration ozone.

In the electric-discharge (or corona) method, an alternating current is imposed across a discharge gap with voltages between 5 and 25 kV and a portion of the oxygen is converted to ozone. A pair of large-area electrodes is separated by a dielectric (1–3 mm in thickness) and an air gap (approximately 3 mm) as shown in Figure 2.15. Although standard frequencies of 50 or 60 cycles are adequate, frequencies as high as 1,000 cycles are also employed.

Only about 10 percent of the input energy is effectively used to produce the ozone. Inefficiencies arise primarily from heat production and, to a lesser extent, from light and sound. Since ozone decomposition is highly temperature dependent, efficient heat-removal techniques are essential to the proper operation of the generator.

The mechanism for ozone generation is the excitation and acceleration of stray electrons within the high-voltage field. The alternating current causes the electrons to be attracted first to one electrode and then the other. As the electrons attain sufficient velocity, they become capable of splitting some O_2 molecules into free radical oxygen atoms. These atoms may then combine with O_2 molecules to form O_3.

Under optimum operating conditions (efficient heat removal and proper

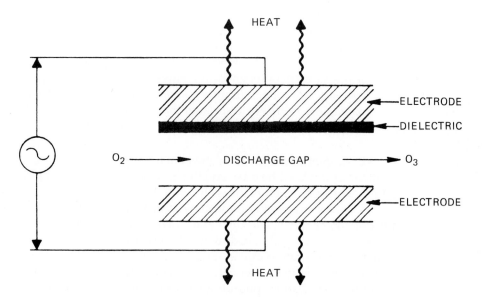

Figure 2.15 Basic ozonator configuration.

feedgas flow), the production of ozone in corona-discharge generators is represented by the following relationships, showing the factors to be considered in the design of these generators:

$$V \alpha\ pg$$

$$(Y/A)\alpha\ \frac{f\epsilon V^2}{d}$$

where:

Y/A = ozone yield per unit area of electrode surface
V = applied voltage
p = gas pressure in the discharge gap
g = discharge-gap width
f = frequency of applied voltage
ϵ = dielectric constant
d = thickness of the dielectric

The following requirements will facilitate optimization of the ozone yield:

- The pressure/gap combination should be constructed so the voltage can be kept relatively low while maintaining reasonable operating pressures. Low voltage protects the dielectric and electrode surfaces. Operating pressures of 10–15 pounds per square inch gauge (psig) are applicable to many waste treatment uses.
- For high-yield efficiency, a thin dielectric with a high-dielectric constant should be used. Glass is the most practical material. High-dielectric strength is required to minimize puncture, while minimal thickness maximizes yield and facilitates heat removal.
- For reduced maintenance problems and prolonged equipment life, high-frequency alternating current should be used. High frequency is less damaging to dielectric surfaces than high voltage.
- Heat removal should be as efficient as possible.

Basic configurations of ozone generators are shown in Figures 2.15 and 2.16: the Otto plate, the tube, and the Lowther plate. Operating characteristics and power requirements for each configuration are given in Table 2.12.

The least efficient of these generators is the Otto plate, developed at the turn of the century. The tube and Lowther plate units include modern innovations in material and design. The Lowther plate generator is the most efficient configuration due in large measure to advantages in heat removal.

In addition to ozone yield, the concentration of ozone is an important consideration. Ozone concentration from a generator is usually regulated by adjusting the flow rate of the feedgas and/or voltage across the electrodes.

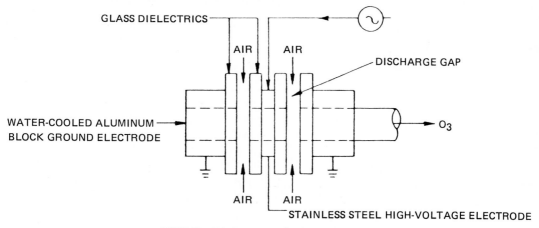

OTTO PLATE-TYPE GENERATOR UNIT

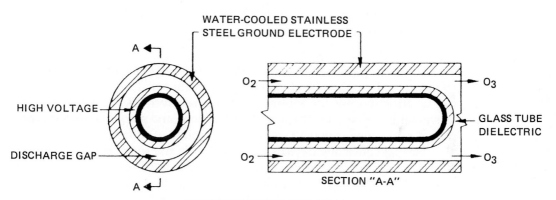

TUBE-TYPE GENERATOR UNIT

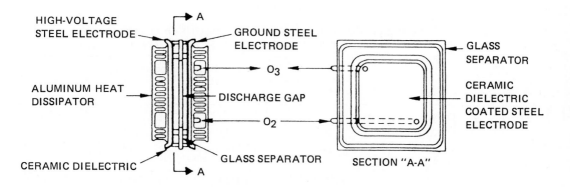

LOWTHER PLATE GENERATOR UNIT

Figure 2.16 Types of ozone generators.

TABLE 2.12   COMPARISON OF COMMERCIAL OZONATORS

Typical Ozonator Operating Characteristics	Type		
	Otto	Tube	Lowther
Feed	air	air, oxygen	oxygen
Dew Point of Feed (°F)	−60	−60	−40
Cooling	water	water	air
Pressure	~0	3–15	1–12
Discharge Gap (in.)	0.125	0.10	0.05
Voltage (kV Peak)	7.5–20	15–19	8–10
Frequency (Hz)	50–500	60	2,000
Dielectric Thickness (in.)	0.12–0.19	0.10	0.02
Power Requirements[a]			
Air Feed	10.2	7.5–10.0	6.3–8.8
Oxygen Feed	—	3.75–5.0	2.5–3.5

[a] kWh/lb of ozone at 1 percent concentration.

Contactor design is important in order to maximize the ozone-transfer efficiency and to minimize the net cost for treatment. The three major obstacles to efficient ozone utilization are ozone's relatively low solubility in water as shown in Table 2.13, the low concentrations and amounts of ozone produced from ozone generators, and the instability of ozone.

Several contacting devices are currently in use including positive-pressure injectors, diffusers, and venturi units. Specific contact systems must be designed for each different application of ozone to wastewater. Further development in this area of gas-liquid contacting needs to be done despite its importance in waste treatment applications.

TABLE 2.13   COMPARISON OF THE SOLUBILITIES OF OZONE, CHLORINE, AND OXYGEN BY WATER TEMPERATURE AND GAS CONCENTRATION

Gas	Solubility By Water Temperature (mg/l)			
	0°C	10°C	20°C	30°C
Oxygen				
@ 100%	70.5	54.9	44.9	38.2
@ 21%	14.8	11.5	9.4	8.0
Ozone				
@ 100%	1,374.3	1,114.9	789.0	499.6
21%	55.0	44.6	31.6	20.0
Chlorine				
@ 100%	14,816.5	9,963.4	7,263.6	5,688.8
@ 99.8%	14,879.4	9,943.5	7,249.1	5,677.4

In order to define the appropriate contactor, the following should be specified:

- The objective: disinfection biochemical oxygen demand (BOD) or chemical oxygen demand (COD) reduction to a particular level, trace refractory organics oxidation, and so on.
- Relative rates of competitive reactions: chemical oxidation, lysing bacteria, decomposition of ozone in aqueous solutions, and so on.
- Mass-transfer rate of ozone into solution.
- Wastewater quality characteristics: total suspended solids, organic loading, and so on.
- Operating pressure of the system.
- Ozone concentration utilized.

Other considerations for the contacting system itself include contactor type (for example, packed bed, sparged column); number and configuration of contactor stages; points of gas-liquid contact, whether the mix is cocurrent or countercurrent; and the construction materials used. It is clear that designing an ozonation system for even a relatively simple application requires a thorough understanding of many factors in order to employ sound engineering methods and optimization techniques.

Applications

Market areas of interest to manufacturers of ozone systems, actual uses defined as those which have been in operation for some time and not including "pilot" studies, arise in the following categories: odor control (sewage treatment and industrial), industrial chemicals synthesis, industrial water and wastewater treatment, and drinking water disinfection.

Ozone has proven to be effective against viruses. France has adopted a standard for the use of ozone to inactivate viruses. When an ozone residual of 0.4 mg/l can be measured 4 minutes after the initial ozone demand has been met, viral inactivation is satisfied. This property plus ozone's freedom from residual formation are important considerations in the public health aspects of ozonation. When ozonation is combined with activated carbon filtration, a high degree of organic removal can be achieved. Concerning the toxicity of oxidation products of ozone and the removal of specific compounds via ozonation, available evidence does not indicate any major health hazards associated with the use of ozone in wastewater treatment.

Odor Control

In the United States, the largest existing market for ozone systems is odor control. Much of this market is in sewage treatment plants. Industrial markets for ozone in odor control are smaller than for sewage treatment plants. Established applications include cooking odors at restaurants; pharmaceutical fermentations; fish, meat, and food processing; plastics and rubber processing; paint and varnish manufacture; and rendering plants. Nearly all of these industrial odor control applications use less than 100 lb/day of ozone and most use between 1–25 lb/day.

Industrial Chemical Synthesis

There has been only one major use for ozone today in the field of chemical synthesis: the ozonation of oleic acid to produce azelaic acid. Oleic acid is obtained from either tallow, a by-product of meat-packing plants, or from tall oil, a by-product of making paper from wood. Oleic acid is dissolved in about half its weight of pelargonic acid and is ozonized continuously in a reactor with approximately 2 percent ozone in oxygen; it is oxidized for several hours. The pelargonic and azelaic acids are recovered by vacuum distillation. The acids are then esterified to yield a plasticizer for vinyl compounds or for the production of lubricants. Azelaic acid is also a starting material in the production of a nylon type of polymer.

Industrial Water and Wastewater Treatment

The markets for ozone in industrial water and wastewater treatment are quite small. Industrial applications for ozone could grow. The use of ozone for treating photoprocessing solutions is a novel application that has been limited, but might grow. In this process, silver is recovered electrolytically; then the spent bleach baths of iron ferrocyanide complexes are ozonated. Iron cyanide complexes are stable to ozonation so that the ferrous iron is merely oxidized to ferric, which is its original form. Thus, the bleach is "regenerated" and is ready for recycling and reuse by the photoprocessor.

The relationship for the oxidation of ferrocyanide to ferricyanide is:

$$2Na_4Fe(CN)_6 \cdot IOH_2O + O_3 \quad 2Na_3Fe(CN)_6 + O_2 + 9H_2O + 2NaOH \tag{2.27}$$

Ideally, 20.2 pounds of the ferrocyanide can be converted to 11.7 pounds of ferricyanide by one pound of ozone. Indeed, ozone oxidation efficiency is nearly 100 percent for ferrocyanide concentrations above 1.0 g/l.

A typical ozone system consists of 100 g/hr at a concentration of 1.0 percent to 1.5 percent in air fed to the bottom of bleach collection tanks through ceramic spargers (pore size of approximately 100 μ). The system contains air compression and drying equipment, automatic control features, and a flat-plate, air-cooled ozone generator. Regeneration of bleach wastes totaling about 10,000 gallons a year, and recovery of other chemicals can also be cost effective.

Municipal Drinking Water

In the United States, Whiting, Indiana and Strasburg, Pennsylvania have used ozone in their drinking water treatment process. Other cities have run pilot studies.

Ozone is used as a bleaching agent for miscellaneous items: petroleum, clays, wood products, and chemical baths. It has been proposed as a bleaching agent for hair and as a disinfectant for oils and emulsions. Ozone is used to modify tryptophan and indigo plant juice. It is an important factor in color-fastness.

The desulfurization of flue gases by ozone has been considered an application where it promotes liquid-phase oxidation. The operations are carried out with vanadium catalysts, and the oxidation step is performed in gas-fluidized beds. The desulfurizing effect of ozone on light petroleum distillates has also been reported.

The use of ozone has been proposed in special ore-flotation processes. Two widely different applications involve hydraulic cement and the fabrication of coating on insulators.

The metallurgical applications include steel refining, electrochemical processes, and gold recovery. The aggressive reactivity of ozone is evident in the corrosion of stainless steel and in chemical etching. The inhibition of ozone decomposition is accomplished in the presence of SF_6, CCl_2F_2, or CF_3. Metal coatings, paints, and lacquers have been evaluated with respect to ozone resistivity.

Ozone has been examined as a potential source of high-energy oxidation and for combustion and propulsion applications.

Ozonation Equipment and Processes

Ozonation systems are comprised of four main parts, including a gas-preparation unit, an electrical power unit, an ozone generator, and a contactor which includes an off-gas treatment stage. Ancillary equipment includes instruments and controls, safety equipment and equipment housing, and structural supports. The four major components of the ozonation process are illustrated in Figure 2.17.

A high level of gas preparation (usually air) is needed before ozone generation. The air must be dried to retard the formation of nitric acid and to increase the efficiency of the generation. Moisture accelerates the decomposition of ozone. Nitric acid is formed when nitrogen combines with moisture in the corona discharge. Since nitric acid will chemically attack the equipment, introduction of moist air into the unit must be avoided.

Selection of the air-preparation system depends on the type of contact system chosen. The gas-preparation system will, however, normally include refrigerant gas cooling and desiccant drying to a minimum dew point of $-40°C$. A dew-point monitor or hygrometer is an essential part of any air-

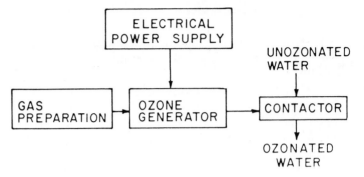

Figure 2.17 Major components of the ozonation process.

preparation unit. Figure 2.18 is a schematic of a low-pressure air-preparation system with turbine contacting.

Conversion efficiencies can be greatly increased with the use of oxygen. However, the use of high-purity oxygen for ozone generation for disinfection is cost effective. The Duisburg plant and the Tailfen plant of Brussels, Belgium, are the only operational municipal water treatment plants known which use high-purity oxygen instead of air as the ozone generator feedgas.

Electrical power supply units vary considerably among manufacturers. Power consumption and ozone-generation capacity are proportional to both

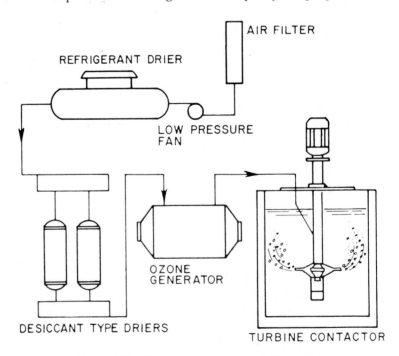

Figure 2.18 A low-pressure gas-preparation system.

voltage and frequency. There are two methods to control the output of an ozone generator: vary voltage or vary frequency. Three common electrical power supply configurations are used in commercial equipment:

- Low frequency (60 Hz), variable voltage.
- Medium frequency (600 Hz), variable voltage.
- Fixed voltage, variable frequency.

The most frequently used is the constant low-frequency, variable-voltage configuration. For larger systems, the 600-Hz fixed frequency is often employed as it provides double ozone production with no increase in ozone generator size.

The electrical (corona) discharge method is considered to be the only practical technique for generating ozone in plant-scale quantities. In principle, an ozone generator consists of a pair of electrodes separated by a gas space and a layer of glass insulator. An oxygen-rich gas is passed through the empty space and a high-voltage alternating current is applied. A corona discharge takes place across the gas space and ozone is generated when a portion of the oxygen is ionized and then becomes associated with nonionized oxygen molecules.

Figure 2.19 shows the details of a typical horizontal tube-type ozone generator. This unit is preferred for larger systems. Water-cooled plate units are often used in smaller operations. However, these require considerably more floor space per unit of output than the tube-type units. The air-cooled Lowther plate type is a relatively new design. It has the potential for simplifying the use of ozone-generating equipment. However, it has had only limited operating experience in water treatment facilities.

After the ozone has been generated, it is mixed with the water stream being treated in a device called a contactor. The objective of this operation is to maximize the dissolution of ozone into the water at the lowest power expenditure. There is a variety of ozone contactor designs. Principal ones employed in wastewater treatment facilities include:

- Multistage porous diffuser contactors, which involve a single application of an ozone-rich gas stream and application of fresh ozone gas to second and subsequent stages with off-gases recycled to the first stage.
- Eductor-induced, ozone vacuum injector contactors, which include total or partial plant flow through the eductor; and subsequent stages with off-gases recycled to the first stage.
- Turbine contactors, which involve positive or negative pressure to the turbine.
- Packed-bed contactors, which include concurrent or countercurrent water/ozone-rich gas flow.

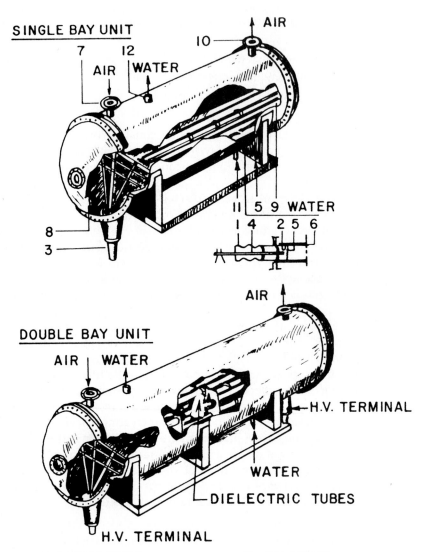

SINGLE BAY UNIT

DOUBLE BAY UNIT

1. DIELECTRIC TUBE
2. METALLIC COATING
3. H.V. TERMINAL
4. CONTACT
5. CENTERING PIECE
6. IONIZATION GAP

7. AIR INLET
8. FRONT CHAMBER
9. REAR CHAMBER
10. AIR OUTLET
11. WATER INLET
12. WATER OUTLET

Figure 2.19   Details of horizontal tube-type ozone generator.

- Two-level diffuser contactors, which involve application of ozone-rich gas to the lower chamber. Lower chamber off-gases are applied to the upper chamber.

Off-gas treatment from contactors is an important consideration. Methods employed for off-gas treatment include dilution, destruction via granular activated carbon, thermal or catalytic destruction, and recycling.

Measurement and Control

Favorable operational economics and good management practices require high levels of control of the ozonation system. Depending on the specific process of ozone applications, plant size, and design philosophy, the control system may be simple or complex. The trend in Europe is toward highly sophisticated and centralized control.

Several parameters should be measured to provide a fully operable ozonation system. There should be a means of providing full temperature and pressure profiles of the ozone generator feedgas from the initial pressurization (by fan, blower, or compressor) to the ozone generator inlet. Moisture content is also important. There should be a means of measuring the moisture content of the feedgas to the ozone generator. This procedure should be conducted with a continuously monitoring dew-point meter or hygrometer.

Other parameters that require monitoring include:

- Temperature, pressure, flow rate, and ozone concentration of the ozone-containing gas being discharged from all the ozone generators. This is the only effective method by which ozone dosage and the ozone production capacity of the ozone generator can be determined.
- Power supplied to the ozone generators. The parameters measured include amperage, voltage, power, and frequency, if this is a controllable variable.
- Flow rate and temperature of the cooling water to all water-cooled ozone generators. Reliable cooling is important to maintain constant ozone production and to protect the dielectrics in the generation equipment.
- There should be a means to monitor the several cycles of the desiccant drier, particularly the thermal-swing unit.

Analytical measurements of ozone concentrations must be made in the ozonized gas from the ozone generator, the contactor off-gases, and the residual ozone level in the ozonized water. Methods of ozone measurement commonly used are the: simple "sniff" test, Draeger-type detector tube, wet chemistry potassium iodide method, amperometric-type instruments, gas-phase chemiluminescence, and ultraviolet radiation adsorption.

The use of control systems based on these measurements varies considerably. The key to successful operation is an accurate and reliable residual ozone analyzer. Continuous residual ozone monitoring equipment may be successfully applied to water that has already received a high level of treatment. However, a more cautious approach must be taken with the application of continuous residual ozone monitoring equipment for water that has only received chemical clarification because the ozone demand has not yet been satisfied and the residual is not as stable.

Ozone production must be closely controlled because excess ozone cannot be stored. Changes in process demand must be responded to rapidly. Ozone production is costly; underozonation may produce undesired effects and overozonation may require additional costs where off-gas destruction is used.

Operation and Maintenance

Ozonation equipment typically has low maintenance requirements. The air-preparation system requires frequent attention for air filter cleaning/changing and for assuring that the desiccant is drying the air properly. However, both are usually simple operations.

Two factors which impact ozone generator operation and maintenance are the effectiveness of the air-preparation system and the amount of time that the generator is required to operate at maximum capacity. Maintenance of the ozone generators is commonly scheduled once a year. However, many plants perform this maintenance every six months. Typically, one man-week is necessary to service an individual ozone generation unit of the horizontal-tube type. Dielectric replacement due to failure as well as breakage during maintenance may be as low as 1 percent to 2 percent. An average tube life of ten years can be expected if a feedgas dew point of $-60°$ is maintained and if the ozone generator is not required to operate for prolonged periods at its rated capacity. Plate-type ozone generators use window glass as dielectrics. However, the same attention to air preparation is taken as with the more expensive glass or ceramic tubes in order to avoid costly downtime.

Operations and maintenance of the ozone contactor also requires attention. Turbines require electricity to power the drive motors, while porous diffusers require regular inspection and maintenance to insure a uniform distribution of ozone-rich gas in the contact chamber. It should be noted that serious safety problems exist with servicing some of these units. For example, even after purging the contact chambers with air, maintenance personnel entering the chambers should be equipped with a self-contained breathing apparatus, since the density of ozone is heavier than air and therefore is difficult to remove completely by air purging.

Potential Applications

The potential application for municipal drinking water applications of ozone is difficult to project because there has been no strong driving force to reduce or eliminate dependence on chlorine in the United States except against potentially carcinogenic-compound generation with the chlorination of water. There has been a great deal of experience with chlorination, and most treatment plant operators have been content to take care of any plant-overloading condition simply by adding more chlorine.

Ozonation has been most widely used for taste and odor control, color removal, and removal of iron and manganese, rather than for disinfection. Most states, in fact, do not recognize any disinfectant for water supplies except chlorine. Standards for tastes, odors, or colors are not addressed by the Safe Drinking Water Act, since these are defined as "aesthetic" contaminants, not "health-related" contaminants.

Municipal Wastewater Treatment—Disinfection

Of the potential uses for ozone in the United States, the treatment of municipal wastewater for disinfection as a replacement for chlorine is of greatest significance. There has been some scientific information showing that residual chlorine and chlorinated products normally found in disinfected sewage treatment plant effluents (chloramines, chlorinated hydrocarbons, and the like) may produce serious detrimental effects on the aquatic organisms found in rivers, streams, and lakes where the disinfected effluents are discharged. Many of the chlorinated organics find their way into the human food chain, beginning with their incorporation into aquatic organisms. The Food and Drug Administration has been concerned with and involved in following chlorinated organics as well as heavy metals through the environment and into the food chain.

Chemical dechlorination, however, makes treatment more complex. During chemical dechlorination, all of the residual oxygen that may have been present as a result of aeration applied during secondary biological sewage treatment will be removed. To meet water quality standards for dissolved oxygen, the dechlorinated effluent now must be reaerated. Thus, the total disinfection process with chlorine may involve chlorination-dechlorination-reaeration, as opposed to the simple chlorination-then-discharge procedures that have been followed for years. By contrast, ozone is as good or better a bactericide and as good or better a viral inactivation agent as chlorine, and its use leaves a high dissolved oxygen residual in the effluent. Thus, the single step of ozonation will accomplish what will require three steps with chlorine.

Tertiary Treatment

Tertiary treatment involves processing after secondary treatment has been effected, and involves reducing excessive biochemical oxygen demand (BOD), chemical oxygen demand (COD), phosphates, ammonia, nitrates, nitrites, and so on. One unique application considered has been the removal of suspended solids with ozone, eliminating the need for chemical precipitants which add to the total dissolved solids (TDS).

Ozone can reduce COD, BOD, ammonia, and nitrites by oxidizing, but only under special conditions not generally required for simple disinfection. Phosphates, nitrates, and dissolved and suspended solids have been reduced during ozonation as a by-product of foaming, caused by the rapid rise of tiny, electrostatically-charged bubbles of ozone-air or ozone-oxygen mixtures through the effluent being ozonated.

Ammonia can be oxidized by ozone, but this occurs rapidly only at high pH. Thus, if a secondary effluent is to receive lime treatment (to remove phosphates), then it would be feasible to treat this stream with ozone prior to lowering the pH for discharging.

COD generally can be reduced with ozone by extending contact times and increasing dosages—both of which add to ozonation costs. Partial ozonation sometimes converts biorefractory COD to BOD, easily degraded biologically. This requires considerably less ozone, although contact times still may be more lengthy than, for example, for disinfection.

Some COD materials which are not readily absorbed by activated carbon can be converted to more easily absorbed materials upon partial ozonation.

In the treatment of municipal wastewaters, ozone would normally be applied following primary and secondary sewage treatment. It would also be an advantage, where possible, to use a filtration stage such as microstraining for tertiary treatment. Water would be virtually free of suspended matter and ozone would be able to act on the dissolved organics and other oxidizable substances.

Summary

Ozonation for drinking water treatment is a well-established and growing technology. Over 100 operational plants throughout the world, mostly in Europe, utilize ozone for one or more purposes. As an oxidant, ozone currently is used to remove colors, taste, odors, algae, organics (phenols, detergents, pesticides, and so on), cyanides, sulfides, iron, manganese, turbidity, to cause flocculation of micropollutants (soluble organics), and to inactivate viruses.

Ozone is also used as a disinfectant, but seldom in the context of an either-or alternative disinfectant to chlorine. It is normal to follow ozone as the primary disinfectant with a small dosage (up to 0.6 mg/l) of chlorine or chlorine dioxide, which provides a residual for distribution systems.

In certain cases, ozonation can be employed as the sole disinfectant. In addition to the obvious requirements (for example, ozonation is the terminal treatment step and the distribution system must be free of contamination), the distribution system should be short and the resistance time of treated water in the system should also be short. Ammonia should not be present and dissolved organic carbon should be less than 0.2 mg/l. It is also advantageous that the temperature of treated water be low to reduce the potential for bacterial regrowth. Organic oxidation products formed upon ozonation are nonhalogenated, are more biodegradable than before oxidation, and are usually less toxic. However, some pesticides pass through intermediate stages of oxidation to produce more toxic materials. Formation of the same or similar nonhalogenated, more toxic intermediates can also occur with the use of oxidants other than ozone; for example, chlorine and chlorine dioxide. Consequently, it is important when using any oxidant for water treatment to know the identity of dissolved organic matter present, the chemistry of the intermediate oxidation stages, and to design sufficient oxidant into the process to guarantee that such intermediate stages are passed and potentially toxic intermediates are destroyed through oxidation.

3

Thermal Treatment and Recovery Methods

THERMAL DISINFECTION

The application of heat is the oldest concentration, disinfection, and sterilization technique. Variations in chemical impurities in water do not interfere with the disinfecting efficiency of heat as they do with chlorine and other chemical disinfectants. Capital costs for necessary equipment with automatic controls are low. However, depending on the nature of the application, this can be an energy-intensive operation with associated high operating costs. In addition to bacterial pathogens, heat readily destroys a variety of microorganisms that are highly resistant to chemical disinfectants. These organisms include amoebic cysts, worms, and viruses.

Application

Dry or wet heat can be used to destroy bacteria. These forms of treatment can be used for air and surface disinfection, respectively. Thermally-induced death of bacteria cells and spores appears to be an exponential function. The death rate of bacteria subjected to heat is logarithmic; that is, the reduction in the number of viable cells is an exponential function of the time exposure at a constant temperature.

The time needed to destroy an unspecified number of organisms in a given time at constant temperature depends on the number of organisms subjected to the heat treatment. Death rate, then, is a function of the population size of cells directly exposed to heat.

Since the death rate has been described as logarithmic, a straight line can be obtained by plotting the logarithm of the number of surviving cells versus exposure time. The equation for this line is the following:

$$t = \theta (\log N_o - \log N_t) \tag{3.1}$$

where

t = time of exposure to the lethal agent
θ = time required to reduce the viable population by 90 percent
N_o = number of viable cells in the initial population
N_t = number of viable cells in the population after exposure time t

The term θ in Equation 3.1 represents the reciprocal of the slope of the survival curve. It can be used for comparing the resistance of different organisms to a given set of lethal conditions.

Note that N_t actually represents the probability of survival and consequently $\theta < N_t < 1$. As an example, if $N_t = 0.01$, this means there is one chance in 100 that an organism in the exposed population will survive.

For temperature-dependent disinfection processes where attainment of a given lethal temperature is not instantaneous (and where cooling from this temperature is not instantaneous), an infinite number of temperatures exists during the heating and cooling phases that may have significant lethal effects. To obtain a true sterilizing value for the total process in any particular case, it is necessary to integrate the lethal effects for temperatures which exist during heating and cooling along with that occurring during any holding period at the maximum temperature. This calls for some resistance parameter to account for the relative resistance of the microbial population to the lethal agent at the different influencing temperatures.

A semilogarithmic plot of θ versus temperature (T) produces a straight line, the equation of which follows:

$$\log \theta_2 - \log \theta_1 = \frac{1}{Z} (T_1 - T_2) \tag{3.2}$$

where

θ_1 = θ value characterizing the resistance of the organism at temperature T_1
θ_2 = θ value characterizing the resistance of the organism at temperature T_2
Z = number of temperature degrees for the straight line to traverse one log cycle (that is, it is the reciprocal of the slope of the line)

This straight line is termed the *thermal destruction curve*, provided that only heat is the lethal agent. If the lethal agent is a chemical in the vapor phase (the activity of which is temperature dependent), the curve is called a *thermochemical destruction curve.* In either case, Z accounts for the relative re-sistance of an organism at all different influencing temperatures. Therefore, θ and Z are the only parameters needed to account for microbial resistance in the mathematical evaluation of any temperature-dependent disinfection process when death of the organism of concern is logarithmic. Values for θ and Z characterizing the resistance of an organism in any temperature-de-pendent system may be determined experimentally with specially designed equipment such as that for moist heat. Once these parameters have been evaluated, Equations 3.1 and 3.2 can be directly applied in the evaluation of disinfection processes.

Activation, Inactivation, and Dormancy

Deviation from the linear dependency of the survival curve takes on two characteristic profiles. First, a hump or lag may appear in the initial portion of the survival curve. This will generally occur with a heat-resistant popu-lation of spores and is referred to as heat activation. The second nonlinearity takes the form of a tail on the final portion of the curve. This is attributed to more than one thermoresistant variant in the population.

An activation energy is generally necessary to initiate a chemical or bi-ological process. In this case, it is the energy necessary to release spores from their dormant state so that they can begin germination. There is also an ac-tivation energy requirement to initiate inactivation (that is, the destruction of microorganisms).

The activation portion of the survival curve of spores is not generally observed in the survival curves of vegetative cells. If such an anomaly exists, it could be explained on the basis of other factors, such as thermal lag or protective effects for vegetative cells, which are not always as predominant as the activation principle for spores. Where such determinations are carefully conducted, thermal lags can be minimized. The activation to germinate por-tions of a survival curve of a population of spores becomes more or less pronounced under the following circumstances: The more heat resistant a strain, the more likely it is to have a pronounced activation requirement. The survival curve of spores of the thermophile Bacillus stearo-thennophilus is shown in Figure 3.1. Note that the curve shows both an initial lag and a pronounced hump.

Treatment Temperature

The lower the treatment temperature, the more pronounced the initial phase is likely to be because of a relatively slow activation rate. Conversely, as treatment temperatures increase, the activation time reduces, the hump be-comes larger, and the lag time shorter (eventually approaching extinction because of the inability to observe the effect over very short increments of

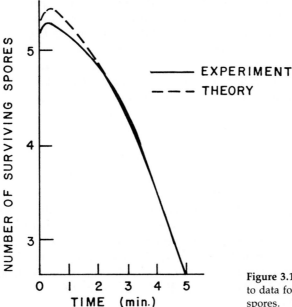

Figure 3.1 Comparison of predictions to data for *Bacillus stearothermophilus* spores.

time). This may partially explain the nonlinearity that exists at the initial portion of survival curves.

The degree of spontaneous germination and the contribution to germination due to constituents of the growth media influence the shape of the initial phase of the survival curve as determined by colony count.

The incubation temperature also contributes to the activation of a fraction of the spores in the population. This is not, however, as great a contribution.

Drying or wetting, storage temperature, and freeze-thaw cycles all contribute to the shape of the survival curve. Other factors such as clumping, diffusion, and thermal lags may also influence the response to heat treatment and the ultimate shape of the survival curves. These factors, however, do not affect the characteristic activation principle which is inherent to the heat-resistant spore.

Activation/Inactivation

An assumption often made in applying thermal disinfection is that all the spores comprising a homogeneous population are dormant and that the growth media contain small amounts of activation chemicals that are sufficient to activate a low percentage of the spores. The incubation temperature will also activate a small percentage of spores. This accounts for the small fraction of spores which always seem to germinate spontaneously and grow to visible colonies.

Heat-induced dormancy for *B. stearothermophilus* spores in a temperature range of 80–100°C, is below what is termed the true activation temperature. This is difficult to demonstrate as the activation rate is generally very slow and matches the inactivation rate. The nonlinear shape of the survival curves for *B. stearothermophilus* spores (and spores of other heat-resistant species) indicates that activation must precede inactivation. The inactivation of dormant spores must proceed at extremely slow rates in comparison to activated spores.

Thermal Resistance

Heat resistance varies among microbial species, proteins, and bacterial spores. Many bacteria spore species are highly resistant to most chemical, physical killing agents, as well as heat. There is a significant difference between the thermal resistances to moist and dry heat. A spore species resistant to one form of heat is not necessarily resistant to the other. A comparison of the spores of thermophile *B. stearothemlophilus* with mesophile *B. subtilus var. niger* is a good example. These spores are used to monitor sterilization processes for moist and dry heat, respectively. To activate 10^5 spores of *B. stearothermophilus* (in saturated steam at 121°C) requires roughly 12 minutes of exposure. However, 10^6 spores of *B. subtilus* are inactivated under identical conditions.

Bacterial death rates are higher in acid or alkaline media than in neutral suspensions. Also, a higher recovery of survivors occurs in the neutral pH zone. Citrate, phthalate, or ammonium buffers reduce thermoresistance of spores compared to those in a phosphate buffer. Spores are more readily inactivated at a low pH, since pH can influence the type of ions that absorb onto the spore surface (this absorption alters the heat stability). Some of the properties of spores resemble the behavior of ion-exchange gels. The spore exchanger resembles a weak cation-exchange system. Consequently, the hydrogen ion possessing the greatest exchange potential would displace other cations. This base-exchange mechanism allows for the adsorption of calcium in excess of the DPA chelation equivalent.

Spore Biosynthesis of Heat-Resistant Components

Sporulation is a multiphase process. First, the refractile cell is formed; then DPA is synthesized. The final step is the development of the thermoresistant spore. Calcium is notably associated with both heat-refractile and relatively water-insoluble compounds. Such an exosporium depleted of calcium and DPA would be expected to act like a porous membrane, which swells under hydrostatic pressure on the absorption of water. Swelling is typical of the germinating spore.

Polypeptides have the capability for extensive intra- and intermolecular bonding of the hydrophobic type. This plays an important role in the composition of spore coats. Also, increased density is produced by the increased binding of heavy atoms, which is presumed to be necessary for sporulation.

Applications

The terms *sterilization* and *disinfection* are often used interchangeably, and this is incorrect. It is only when heat is used to affect microorganisms that both have the same meaning. Probably the oldest disinfection application is that of boiling water. Most water can be sterilized by simply boiling for 10 to 20 minutes. The application of thermal treatment for disinfection is not limited to water treatment. There are numerous applications in the canning and food industries as well. There are three specific applications in which thermal treatment is widely used. These are in the processing of dry goods, pasteurization of water, and the thermal conditioning of sludges.

The time required to destroy *B. subtilis var. niger* at 121°C is almost 2,000 times as long in hot air as it is in steam. Saturated steam at 121°C provides at least seven times more available heat as air. The time needed to kill organisms is not directly proportional to the amount of available heat, but rather depends on different reaction mechanisms which are related to steam versus dry heat.

Steam sterilization is a highly effective method. There are, however, difficulties in its application, namely, moisture penetration, air removal and entrainment, heat and moisture damage, and wetting.

To obtain efficient air removal, it is necessary to use low-temperature steam in conjunction with a high vacuum. A high-vacuum unit will provide a small but steady influx of steam to the sterilization chamber during evacuation. This type of system is prone to air entrainment, which can result from three different causes, namely, a small load in a large-volume apparatus, insufficient removal of air by prevacuum, and air leakage into the system.

Air having low transitional kinetic energy is driven by relatively high-energy steam. If the air is well distributed in a large enough load, no air will be entrained (assuming no air leaks and if the degree of prevacuum is sufficient). The severity of the air-entrainment problem depends on the amount of it available relative to the load and its energy state. The greater the energy differential, the worse the air entrainment will be.

The advantages of steam processing dry goods include a fast cycle time where much larger volumes of material can be processed per day and the fact that damage to materials is minimized because relatively short exposure times are needed.

The success achieved in killing pathogens in milk via pasteurization suggested the use of heat as a means of destroying coliform organisms and eliminating the problems of chemical feeding. Variations in chemical impurities in water do not interfere with the disinfection efficiency of heat as they do with chlorine and many other chemical disinfectants. In addition to bacterial pathogens, heat readily destroys other organisms, such as cysts, worms, and viruses. These are generally more resistant to chemical disinfectants than vegetative bacteria.

Technology for disinfecting small quantities of water with heat is well established. However, it is difficult, time consuming, and troublesome to disinfect large quantities of water. Efforts have been directed at reducing the required exposure time to achieve sterilization. These efforts have been made in both the heating and cooling phases. The water being treated must be heated to a high temperature very rapidly, followed by a rapid cooling cycle. This, however, requires careful control.

The same time-temperature relationships hold for water and water solutions as for dry goods in steam. However, the time necessary to attain a specified temperature must be added to the process. The dairy and canning industries have used automated processes integrating time-temperature requirements over the heating and cooling phase of the treatment.

For water pasteurization to be economical, it is necessary to use the heat of the treated water to warm the water to be treated. Also, such a system must be operated at a nearly steady flow rate to reduce heat losses. This requires proper surge tanks and reservoirs from which treated water can be recycled for treatment of incoming streams.

In a typical treatment system, untreated water would enter a heat exchanger. An electric heater would raise the water temperature to the pasteurization temperature. A 15-second retention time at the pasteurization temperature is typically needed. This would be done in a retention tube. An air-relief valve would permit escape of gases liberated at the higher temperature. One of the drawbacks with such a system is that water temperatures can fall below 161°F. To prevent this, a solenoid valve and a thermostat switch must be properly arranged. When the water is hotter than 161°F, the water flows through the heat exchanger, where it supplies its heat to the incoming water and is then cooled. The heat exchanger is generally the most expensive component in this system.

For proper evaluation of such a system, the following areas should be evaluated:

- The effectiveness of the unit in killing bacteria.
- The power consumption required for the unit.
- Maintenance requirements and reliability of the equipment.
- The significance of scale or corrosion in reducing the operating efficiency or life of the equipment.

THERMAL CONDITIONING OF SLUDGE

Wastewater sludge disinfection, the destruction or inactivation of pathogenic organisms in the sludge, is carried out principally to minimize public health concerns. Destruction is the physical disruption or disintegration of a pathogenic organism, while inactivation can be defined as the removal of a pathogen's ability to infect. Presently in the United States procedures to reduce the number of pathogenic organisms is a requirement before sale of sludge or sludge-containing products to the public as a soil conditioner, or before recycling sludge to croplands. Since the final use or disposal of sludge may differ greatly with respect to health concerns, and since a great number of treatment options effecting various degrees of pathogen reduction are available, the system chosen for the reduction of pathogens should be tailored to the specific application.

Thermal conditioning of sludge in a closed, pressurized system destroys pathogenic organisms and permits dewatering. The product generally has a good heating value or can be used for land filling or fertilizer base. In this process, sewage sludge is ground and pumped through a heat exchanger and sent with air to a reactor where it is heated to a temperature of 350–400°F. The processed sludge and air are returned through the heat exchanger to recover heat. The conditioned slurry is then discharged to a gravity thickener where the vapors are separated and the solids are concentrated (thickened). The treatment process renders the sludge easily dewaterable without the addition of chemicals.

After thickening, a variety of sludge handling and disposal options are available. For example, the thickened sludge can be applied directly to land. If liquid disposal is not applicable to a specific project, the thickened sludge can be dewatered by centrifugation, vacuum filtration, or filter pressing. The dewatered residue can then be land fill or incinerated.

Thermal sludge conditioning and its effects on the chemical and physical structures of wastewater sludge can be best understood from analyses of typical sewage sludges. Wastewater sludge is a complex mixture of waste solids forming a gelatinous mass that is nearly impossible to dewater without further treatment. The organic fraction of the sludge consists of lipids, proteins, and carbohydrates, all bound by physical-chemical forces in a predominantly water-gel-like structure. When the sludge is heated under pressure to temperatures above 350°F (176.5°C), the gel-like structure of sludge is destroyed, liberating the bound water. Dewatering by filtration without chemical conditioning is then a simple matter.

There are several characteristics of thermally conditioned sludges which have an important effect on the cost of plants utilizing thermoconditioning. Various factors, such as thickening properties, dewatering properties, heavy metal distribution, heating value, volatile solids solubilization, and others, all have a major impact on the evaluation of various process alternatives and ultimate disposal.

Thickening and dewatering properties vary depending on the type of sludge. In general, vacuum filtration rates vary from 2–15 lb/ft^2/hr, with cake moistures ranging from 50–70 pc. Lower values (2–4/ft^2/hr) are observed for high proportions of waste-activated sludge; the higher values (up to 15 lb/ft^2/hr) are observed for sludges which are predominantly primary sludge.

Similar results have been obtained for filter pressing. Mixtures of primary and waste-activated sludge of relatively the same proportions which have been thermally conditioned and dewatered at rates from 2,540 lbs dry solids per ft^3/hr, with cake moistures ranging from 50–60 pc.

Heating values of thermally conditioned sludge cake are typically about 12,500 Btu/lb. No marked differences in heating values have been found for different types of sludges. Sludges conditioned with ferric chloride and lime have been found to have heating values in the range of 9,000–10,000 Btu/lb. The lower values experienced for chemically conditioned sludges could be due to:

- Selective solubilization of materials of lower heating value in the thermal conditioning process.
- Endothermic reactions with the conditioning materials.
- Operational differences in the analytical methods used for determining the heating value of the volatile content.

Solubilization of a fraction of the influent-suspended solids can occur as a result of thermal conditioning. In low-pressure, wet-air oxidation, some of the organics present are oxidized as well. Solubilization of the volatile suspended solids produces a supernatant or filtrate of relatively high organic strength.

Ash solubilization and volatile suspended-solids oxidation also decrease the solids loads to downstream solids-handling units.

TABLE 3.1 PATHOGENIC HUMAN BACTERIA POTENTIALLY IN WASTEWATER SLUDGE

Species	Disease
Arizona hinshawii	Arizona infection
Bacillus cereus	*B. cereus* gastroenteritis; food poisoning
Vibrio cholerae	Cholera
Clostridium perfringens	*C. perfringens* gastroenteritis; food poisoning
Clostridium tetani	Tetanus
Escherichia coli	Enteropathogenic *E. coli* infection; acute diarrhea
Leptospira sp	Leptospirosis; Swineherd's disease
Mycobacterium tuberculosis	Tuberculosis
Salmonella paratyphi, A, B, C	Paratyphoid fever
Salmonella sendai	Paratyphoid fever
Salmonella sp (over 1,500 serotypes)	Salmonellosis; acute diarrhea
Salmonella typhi	Typhoid fever
Shigella sp	Shigellosis; bacillary dysentery; acute diarrhea
Yersinia enterocolitica	Yersinia gastroenteritis
Yersinia pseudotuberculosis	Mesenteric lymphadenopathy

There are several advantages that thermoconditioning has over chemical conditioning. These include the sterility of the end product and a residue that can be readily thickened. Bacteria are numerous in the human digestive tract; humans excrete up to 10^{13} coliform and 10^{16} other bacteria in their feces every day. The most important of the pathogenic bacteria are listed in Table 3.1, together with the diseases they cause which may be present in municipal wastewater treatment sludges. Table 3.2 lists potential parasites in wastewater sludge.

Sludge stabilization processes are ideally intended to reduce putrescibility, decrease mass, and improve treatment characteristics such as dewaterability. Many stabilization processes also accomplish substantial reductions in pathogen concentration. Sludge digestion is one of the major methods for sludge stabilization. Well-operated digesters can substantially reduce virus and bacteria levels but are less effective against parasitic cysts.

The requirement for pasteurization is that all sludge be held above a predetermined temperature for a minimum time period. Heat transfer can be

TABLE 3.2 PATHOGENIC HUMAN AND ANIMAL PARASITES POTENTIALLY IN WASTEWATER SLUDGE

Species	Disease
Protozoa	
Acanthamoeba sp	Amoebic meningoencephalitis
Balantidium coli	Balantidiasis, Balantidial dysentery
Dientamoeba fragilis	Dientamoeba infection
Entamoeba histolytica	Amoebiasis; amoebic dysentery
Giardia lamblia	Giardiasis
Isospora bella	Coccidiosis
Naegleria fowleri	Amoebic meningoencephalitis
Toxoplasma gordii	Toxoplasmosis
Nematodes	
Ancyclostoma dirodenale	Ancylostomiasis; hookworm disease
Ancyclostoma sp	Cutaneous larva migrans
Ascaris lumbricoides	Ascariasis; roundworm disease; Ascaris pneumonia
Enterobius vermicularis	Oxyuriasis; pinworm disease
Necator americanus	Necatoriasis; hookworm disease
Strongyloides stercoralis	Strongyloidiasis; hookworm disease
Toxocara canis	Dog roundworm disease, visceral larva migrans
Toxocara cati	Cat roundworm disease; visceral larva migrans
Trichusis trichiura	Trichuriasis; whipworm disease
Helminths	
Diphyllobothrium latum	Fish tapeworm disease
Echinococcus granulosis	Hydatid disease
Echinococcus multilocularis	Aleveolar hydatid disease
Hymenolepsis diminuta	Rat tapeworm disease
Tymenolepsis nana	Dwarf tapeworm disease
Taenia saginata	Taeniasis; beef tapeworm disease
Taenia solium	Cysticercosis; pork tapeworm disease

accomplished by steam injection or with external or internal heat exchangers. Steam injection is preferred because heat transfer through the sludge slurry is slow and not dependable. Incomplete mixing will either increase heating time, reduce process effectiveness, or both. Overheating or extra detention are not desirable, however, because trace metal mobilization may be increased, odor problems will be exacerbated, and unneeded energy will be expended. Batch processing is preferable to avoid reinoculations if short circuiting occurs.

The flow scheme for a sludge pasteurization system with a one-stage heat recuperation system is shown in Figure 3.2. System components include a steam boiler, a preheater, a sludge heater, a high-temperature holding tank, blow-off tanks, and storage basins for the untreated and treated sludge.

Pasteurization is used in Europe and is required in Germany and Switzerland before application of sludge to farmlands during the spring-summer growing season. Based on European experience, heat pasteurization is a proven technology, requiring skills such as boiler operation and understanding of high-temperature and pressure processes. Pasteurization can be applied to either untreated or digested sludge with little pretreatment. Digester gas, available in many plants, is an ideal fuel and is usually produced in

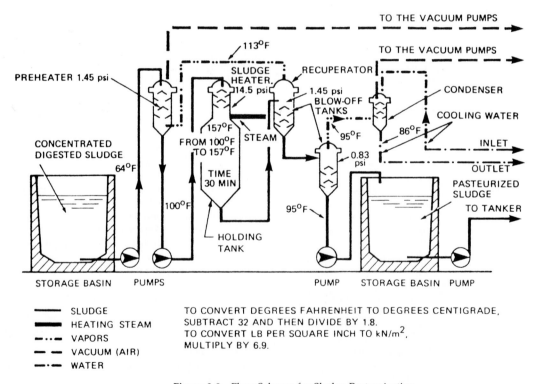

Figure 3.2 Flow Scheme for Sludge Pasteurization.

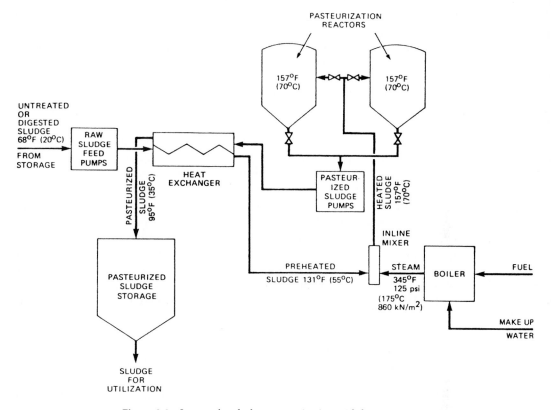

Figure 3.3   Layout for sludge pasteurization with heat recovery.

sufficient quantities to disinfect locally produced sludge. Disadvantages include odor problems and the need for storage facilities following the process—where bacterial pathogens may regrow if sludge is reinoculated.

A pasteurization system should be designed to provide a uniform minimum temperature of 157°F (70°F) for at least 30 minutes. Batch processing is necessary to prevent short circuiting and recontamination, especially by bacteria. In-line mixing of steam and sludge is a possible aid to increase heat-transfer efficiency and assure uniform heating. In-line mixing also eliminates the need to mix the sludge while it is held at the pasteurization temperature.

Figure 3.3 provides a schematic layout for process components of a sludge pasteurization process.

Other Heat Treatment Processes

Heat processes can be classified as heat conditioning, heat drying, high-temperature combustion, and composting. Heat conditioning includes processes where wet wastewater sludge is pressurized with or without oxygen and the temperature is raised to 350°F to 400°F (177°C to 240°C) and held for 15 to 40 minutes.

High-temperature processes include incineration, pyrolysis, or a combination such as starved-air combustion. These processes raise the sludge temperature above 930°F (500°C), destroying the physical structure of all sludge pathogens and effectively sterilizing the sludge. The product of a high-temperature process is sterile unless short circuiting occurs within the process.

Composting is mentioned as a heat process because a major aim of sludge composting operations is to produce a pathogen-free compost by achieving and holding a thermophilic temperature. A well-run composting process greatly reduces the numbers of primary pathogens. However, windrow or aerated pile operations have not achieved a sufficiently uniform internal temperature to inactivate all pathogens. Adverse environmental conditions, particularly heavy rains, can significantly lower composting temperatures. An additional problem with composting is the potential growth of bacteria. Storage of compost for several months following windrow or pile composting helps to further reduce pathogen levels. Secondary pathogens, particularly heat-resistant fungi such as *Aspergillus*, have been found to propagate rapidly during the composting of wastewater sludges. Enclosed mechanical composting systems may achieve sufficient temperature, 157°F (70°C) or greater, for an adequate time.

The use of high-energy radiation for wastewater sludge disinfection has been considered for many years. Beta and gamma rays offer the best potential system performance. Beta rays are high-energy electrons, generated with an accelerator for use in disinfection, while gamma rays are high-energy photons emitted from atomic nuclei. Both types of rays induce secondary ionizations in sludge as they penetrate. Secondary ionizations directly inactivate pathogens and produce oxidizing and reducing compounds that in turn attack pathogens. These methods are discussed elsewhere in the book.

Water Sterilization

Water intravenous fluids, culture media, and many foods are sterilized by steam under pressure in autoclaves or retorts at temperatures above 110°C. Hospitals and food industries usually use a batch process, but in industrial fermentation plants large volumes of culture media are sterilized more satisfactorily by a continuous high-temperature method.

Aqueous fluids provide their own moisture and are heated by conduction through the walls of the container and convection in the fluid itself. They may be dispensed in plugged or sealed containers. In a batch process the fluid is heated in its container until the whole of the contents has reached the temperature selected for sterilization. It is then held for the necessary time and finally cooled. The sterilization cycle is made up of four components: chamber heat-up time, penetration time, holding time, and cooling.

Heat penetration into a fluid is influenced by its volume and viscosity, the wall thickness of the container, and the rate of heating of the autoclave. The viscosity of the fluid may retard the rate of heating by interfering with convection. This is important in the retorting of canned fluids and in the sterilization of agar media.

The rate of heating of an autoclave varies mainly with the size of the load and whether the walls are hot or cold at the beginning of the run. If the chamber reaches the sterilizing temperature quickly (for example, in 5 minutes), the temperature of the fluid will be low at the commencement of the sterilization time. On the other hand, slow heating of the autoclave (for example, 25 minutes) allows the fluid temperature to follow that of the chamber more closely, thus shortening the sterilization time.

In batch sterilization the usual holding times for heat stable solutions are 12 minutes at 121°C or 30 minutes at 115°C. In some cases holding times must be modified to avoid undesirable changes, such as the discoloration of dextrose solutions, which occurs at 115°C, or the loss of gel strength by hydrolysis of agar when autoclaved at an acid pH.

Industrial fermentations are typically carried out in volumes of 5,000 gallons or more. The heating of such large quantities can be best achieved by continuous sterilization of the medium. A continuous sterilizer consists of heating, holding, and cooling sections. The medium is heated through steam-jacketed pipes, by direct injection of steam or by passing over a plate exchange. Cooling is accomplished in water-jacketed coils or in a flask chamber cooler. The important advantage of the continuous process is that heating and cooling of a medium are almost instantaneous. This permits the use of high temperatures up to 150°C (302°F). At the higher temperatures the sporicidal rate is increased to a greater extent than the inactivation of nutrients. Factors other than temperature may influence the exposure time in continuous sterilization. Viscous fluids, for example, exhibit a variable flow rate, the velocity being greatest in the center of the stream, which as a result will not receive as much heating as the periphery. The bacteria carrying particles greater than 1 mm are not easily sterilized by the continuous sterilization method and should be removed by centrifugation or filtration.

Heat Processing of Foods

Heat processing of canned foods is an important commercial application of steam under pressure. Technically known as cooking or retorting, the process is designed to sterilize and cook the food at the same time. It is typically a batch process in which the total amount of heat treatment for sterilization must be compatible with the preservation of flavor, texture, and nutritional quality of the product. Most foods are processed, therefore, to a level of commercial sterility rather than to a state of bacteriological sterility. Though unsterile, they are safe for human consumption and have an acceptable shelf life. The amount of heat processing is carefully calculated to destroy clostridium botulinum spores, which at neutral pH may survive 4 minutes at 120°C, 10 minutes at 115°C, or 33 minutes at 110°C. A high inactivation factor, up to 10^{42}, depending on the initial contamination may be required in the case of nonacid foods.

Heat-penetration time depends on the size and shape of the can and on the physical state of the food, that is, solid or liquid condition. The rapid heating of fluid packs is due to convection, while solid foods heat only by conduction, which is slow. Heat penetration of foods that contain both solid and fluid components, such as baked beans with occasional bits of pork suspended in tomato sauce, can be accelerated by rotating the can during retorting; whether this is done or not depends on the commercial value of product.

The pH of the food is important in determining the amount of heat processing required. Acid foods (pH below 5) such as fruits and pickles, may be safely processed at 100°C, whereas nonacid types (pH 5.0 and over) such as meats, vegetables, and soups must be processed at higher temperatures. The microbial flora of the raw material must be considered in determining heat-processing times and temperatures. The acidity not only increases the sensitivity of bacterial spores to heat but also inhibits their germination. Another method of inducing sensitivity of clostridium spores is by prior exposure to small doses of gamma radiation or by addition of antibiotics.

EVAPORATORS

The task of concentrating a solution by the evaporation of part of the waste or wastewater is one which must be repeatedly performed in some cases. Most frequently the solvent is water; in such instances, the task of the evaporator is the vaporization of part of the water with the resulting production of a concentrated solution. Vapor may be of no use except perhaps for its heat value. It is the residual solution or residue which is saved. A distinction is thus possible in general terms between what is here called the *evaporator*, and a *still*, for in the latter it is the vapor which has value, while generally, though not always, the residual liquor has none. The heating agent is generally steam and, frequently, steam at low pressure (such as 5 to 40 psig).

One of the standard forms of evaporators is an upright cylinder, within which a nest of upright steam tubes is placed. An earlier form has a rectangular base with a dished cover piece, and contains a bundle of horizontal steam tubes. The horizontal steam tubes are also built into the upright cylindrical body. Instead of tubes, suitably placed coils within which steam circulates are installed for small-scale and for batch operation. Certain evaporators have both tubes and at a lower level coils. (See Figure 3.4.)

An example of the upright cylindrical type is the natural circulation short-tube, nonsalting evaporator. It is made in many sizes varying from 2.5–12 feet in diameter, depending on the capacity desired. In these so called

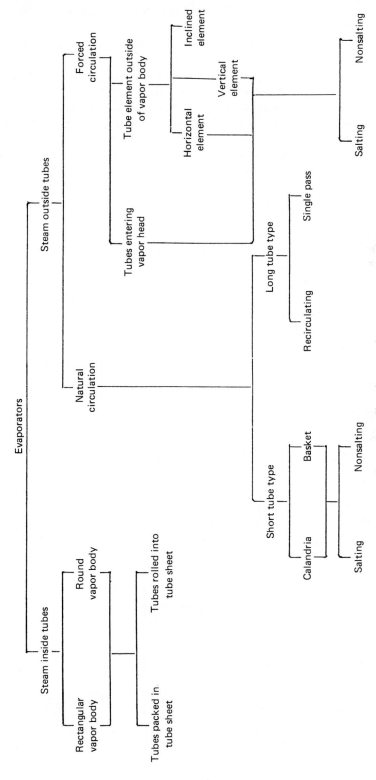

Figure 3.4 Classification of evaporators.

short-tube evaporators, the tubes vary from 2–6 feet in height while their diameter may range from 1–3 inches. In general, the relation of the dimensions of the various parts will be similar to those of a specific example, such as one 8 feet in diameter and 18 feet high in which the tubes are 5 feet long and 2 inches in outside diameter. The tube dimensions given would be averages for those used in the chemical industries. The liquor is inside the tubes, and the steam is on their outer walls. The tubes are set in an upper and lower tube sheet which forms part of the body. A wide central space in the chest, called the downtake, is provided down which this slightly colder liquor travels, while in the tubes, the very hot liquor travels upward. A natural circulation is provided which constantly disturbs the film clinging to the inner wall of the tubes, sweeping it away, and replacing it by fresh liquid—actions which promote evaporation. (See Figure 3.5.) The cross section of the downtake is generally 50 percent of the total cross section of the tubes, but may be more. The steam chamber has an inlet for steam, an outlet for the condensate, and a vent to purge the noncondensable gases. The body has an inlet for the weak liquor, an outlet for the strong liquor, an outlet for the vapors, gauges for pressure and for liquor level, hand holes, manhole, and a generous coat of insulating material to reduce the loss of heat by radiation.

The natural-circulation short-tube salting-out evaporator would be like the one just described, except that its bottom would be made conical; the salt separating as a solid during the concentration which collects by gravity in this cone, from which it may be removed into a salt box or by means of a pump to a point farther away.

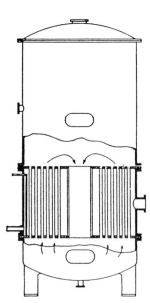

Figure 3.5 A natural-circulation vertical short-tube nonsalting evaporator. The tube sheets are fastened to the body.

CRYSTALLIZERS

A crystallizer consists of a vessel in which a hot solution of proper strength is allowed to cool and form crystals; as a rule, after crystal separation, there remains a mother liquor which may be run off, and with it soluble impurities. (See Figure 3.6.) Crystallizers for inorganic salts are rectangular boxes without covers, made of steel or lined in lead, and provided with an opening for running off the mother liquor. The dimensions vary, but are typically 6 feet by 10 feet, with a depth of 18 or 24 inches. The box is mounted on horses and tilted slightly so that when the run-off plug is removed, the mother liquor may run out to a trough which leads to a collecting vessel. Crystals left in the crystallizer may be shoveled over the edge onto buggies, or small cars on rails, and these are dumped through an opening in the floor to a storage room or shipping room below.

The moisture, or mother liquor, which clings to the crystals is removed first by sending the crystals from the buggy or rail car to centrifugals; from this the comparatively dry crystals are discharged at the bottom, and enter a rotating cylindrical screen to separate them into sizes. A screen with dif-

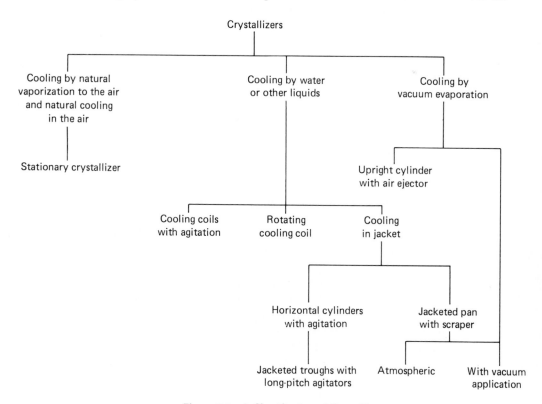

Figure 3.6 A Classification of Crystallizers.

ferent sizes of perforations furnishes various sizes of crystals. The stationary crystallizer is simple and inexpensive. It has, however, at least two defects; one is that it is labor intensive, and the other is that several sizes of crystals are formed, while only one size may be saleable. A third defect is that much time is required for the crop of crystals to deposit.

Jacketed-trough Crystallizer with Agitators

The stationary crystallizer is cooled by natural evaporation and by air cooling of its walls and bottom. More rapid cooling is obtained by using cold water in a jacket, and a better heat transfer is assured if a scraper is used to keep the walls free from crystals. The jacketed crystallizer may then take the form of a trough in which long-pitched ribbon scrapers turn. There is discharged from the trough a slurry of crystals and mother liquor which are sent to the centrifugal; the discharge may be continuous, if hot liquor is fed in at the other end of the trough continuously. Thus, a crystallizer with continuous operation results. The trough can be 24 inches wide by 26 inches deep and 10 feet long; four sections fitted together give a deck 40 feet in length, generally the maximum. If the flow of the liquor must be more rapid than the wall area permits, a second or third deck is provided, and the incomplete crystallized liquor from the first cascaded to the second, and then to the third deck.

Double-Pipe Crystallizer

The double-pipe crystallizer consists of fully jacketed pipes with long-pitch agitators, set up in tiers of three or more, each tier operating in series. The inner pipe may be 24 inches in diameter surrounded by a suitably larger pipe, both roughly 10 feet long; several lengths, generally three, are assembled to form one single horizontal length. Three such lengths are set one above another; the hot liquor fed to the upper one moves along the length of the pipe, drops to the intermediate level, travels through its length, and then drops to the lowest level, where it moves again over its whole length to reach the outlet. The cooling water or brine may be fed countercurrent to the flow of the liquor, or each level may receive fresh cooling medium. If both pipes are cast iron, the outer pipe carries baffles in the annular jacket space which force the water to travel in a spiral path; when the outer pipe is of cast iron and the inner one of steel, stainless steel, or similar metals, a steel rod is welded to the outer side of the latter pipe to form a spiral and produce the same flow in the cooling water. As the liquor moves along the pipes, crystals form, and a slurry is discharged at the outlet which may be handled in one of two ways. It may be sent past a restricting valve to a magma tank at a lower level, or a riser may cause it to reach the level of the upper tier, there to be discharged into a magma tank. The double-pipe condenser shown in Figure 3.7 has such a riser. By using the latter, the velocity of the liquor is made to depend solely on the rate of feed. The agitators are driven by sprockets and chains in the

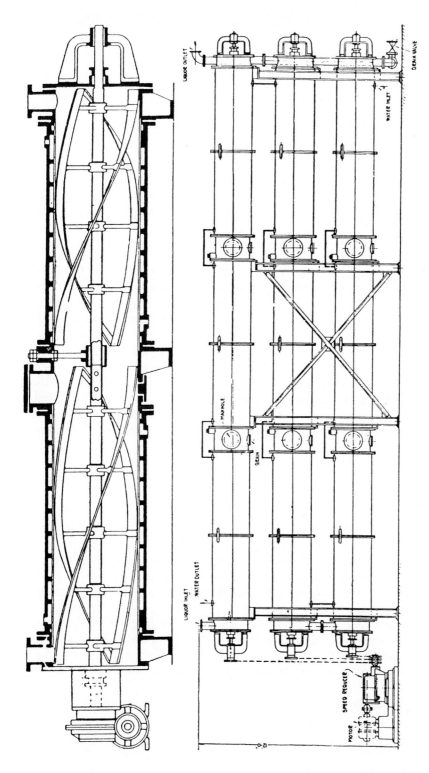

Figure 3.7 (*Upper figure*): Vertical cross-section through one section of a double-pipe crystallizer, showing the scraper-agitator, the adjustable support of the shaft bearing, and the baffles in the cooling jacket. The scraper is adjusted to scrape the bottom of the pipe. (*Lower figure*): A side elevation of a complete double-pipe crystallizer, consisting of two vertical rows of three pipes each, 24 inches outer diameter, 36 feet length, with liquor or slurry outlet brought to the same level as the inlet. The agitators are driven by sprocket chains.

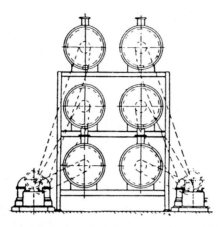

Figure 3.8 Front elevation of the same double-pipe crystallizer shown in side elevation nearby.

assembly shown. The shaft of the agitator is adjustable, and the provision which makes this adjustment easy is visible in Figure 3.8; the adjustment is such that the bottom of the pipe is continuously scraped clean. The rate of rotation of the agitator varies between 5 and 30 rpm, depending on the material handled.

The operation of the crystallizer is continuous; liquor is fed in at the top inlet and slurry is collected from the outlet, both continuously. The crystallizer may readily be made a link in a sequence of operations, all continuous. The double-pipe crystallizer has a large cooling surface, nearly double that provided in the trough crystallizer; the whole circle of the pipe is cooled. The cubical content per linear foot also is greater by 25 percent than that of a trough crystallizer of the same diameter.

	COOLING SURFACE PER LINEAL FOOT	CUBICAL CONTENT PER LINEAL FOOT
Double-pipe crystallizer (24")	6.3 sq. ft.	315 cu. ft.
Trough crystallizer (24")	3.7 sq. ft.	2.57 cu. ft.

Another feature of the double-pipe crystallizer is that the cooling is entirely by the jacket; no evaporation takes place, so that there is no change in the ratio of solute to solvent. The double-pipe crystallizer may be operated under pressure.

The jacketed crystallizer may also have the form of a circular pan in which a scraper agitator slowly rotates. Water applied to the jacket brings about cooling while the crystallization proceeds. The construction of the crystallizer is similar to that of the vacuum-pan drier.

Batch Crystallizers with Coils

A second way to cool the crystallizer with cold water is by means of one or several stationary coils through which the water circulates, while an agitator keeps the liquor in motion, preventing coating and consequent insulation of the coil walls, and bringing new hot layers into contact with the cold walls. The cooling is rapid, and the crystals formed are small and uniform. The resulting slurry remains in the vessel until crystallization is complete, when it is discharged to centrifugals or other devices for separating the crystals from the mother liquor.

An eccentric-coil crystallizer consists of a horizontal cylindrical vessel with a horizontal shaft on which is mounted a cooling coil in a position eccentric to the shaft. (See Figure 3.9.) The latter also carries an agitator scroll by means of diamond-shaped, self-cleaning arms; this rotates with the coil. The cooling water enters and leaves the coil through the hollow shaft. The mass to be cooled is introduced through the charging door at the top and discharged through the opening in the end plate near the bottom. The rotation of the coil is generally slow.

The process of crystallization is more than a matter of cooling a liquor until crystals form. Also, the process of crystallization varies from one substance to another and even for one substance crystallized from various mother liquors. Three phases may be distinguished: the first involves cooling the liquor to the temperature at which the nuclei are about to separate out; the second is the further cooling which leads to the formation of nuclei; the third is the period of growth during which the temperature is held constant. Over the latter stage, the heat evolved by crystallization is removed, but the temperature is not allowed to drop. The importance of observing these phases, and the time to be allowed for each, will depend on the particular process. A relatively pure liquor will require less time for crystal growth than a second

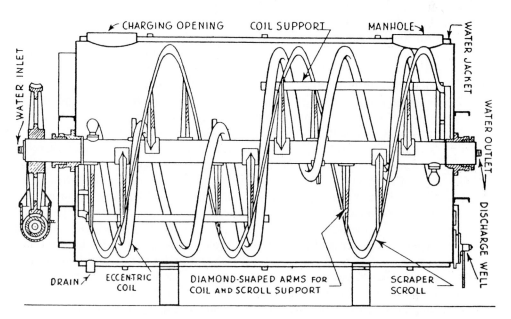

Figure 3.9 Eccentric-coil crystallizer showing the eccentric coil, the diamond-shaped arms, and the scraper scroll.

mother liquor, and the second one in turn will require less time than the third. The time element has a great effect on the yield, for unless time is allowed, the growth of the crystals will be incomplete, and crystallizable material will remain in solution and leave with the mother liquor. In order to reach a high yield, then, a longer time of contact between crystals and mother liquor must be allowed, or the temperature must be lowered. As has been implied already, the amount of impurities and also the pH of the solution have an effect on the yield. These are all reasons why batch crystallization remains of value. Furthermore, batch crystallization is of importance whenever precise control of purity is necessary, for it permits the examination of each batch. In a continuous process, it is difficult to maintain exactly conditions which would confine impurities to the arrow limits sometimes set.

DRYING

Certain materials may be treated directly with fire gases without damage; others are more delicate and must be treated indirectly with heat or warm air. Drying in vacuum permits the maintenance of lower temperatures, but it is more costly than atmospheric drying. Each group of materials has its own class of driers. Liquids containing dissolved solids and thin slurries carrying suspended matter may be handled in drum or spray driers. Sludges and pastes may be mixed with dried materials until they crumble, and this mixture then is dried as solids would be. Wet solids are frequently dried in rotary driers—cylinders lying lengthwise through which a current of heated air or of flue gases travels. Compartment driers are closets or even rooms in which the material is spread on trays or in shallow pans. The operation is discontinuous; in order to make it continuous, the pans may be placed on trucks traveling on rails, and these pushed through a tunnel in which warm air circulates. The belt drier and the chain drier move in the same direction.

Various devices that might meet the needs mentioned and fit other circumstances are shown in Figure 3.10. The process of drying a chemical substance is not simple. In the act of drying a finely divided solid carrying 30 percent to 40 percent water, for example, the rate of evaporation is constant and high as long as the surfaces exposed are wet. After the surface is dry, the water in the interstices must make its way to the surface, a process of diffusion which is slower than evaporation from a wet surface; the rate will then drop. This second part of the process must be modified according to the ease with which the material crumbles as it dries, exposing new surfaces. Drying is concerned with heat transfer and material transfer: heat transfer—such as latent heat of vaporization, and sensible heat to the vapor molecules, to the water in the pores, and to the solid; material transfer—mainly the transfer of water from the inner portions of the paste or crystal mass to the surface.

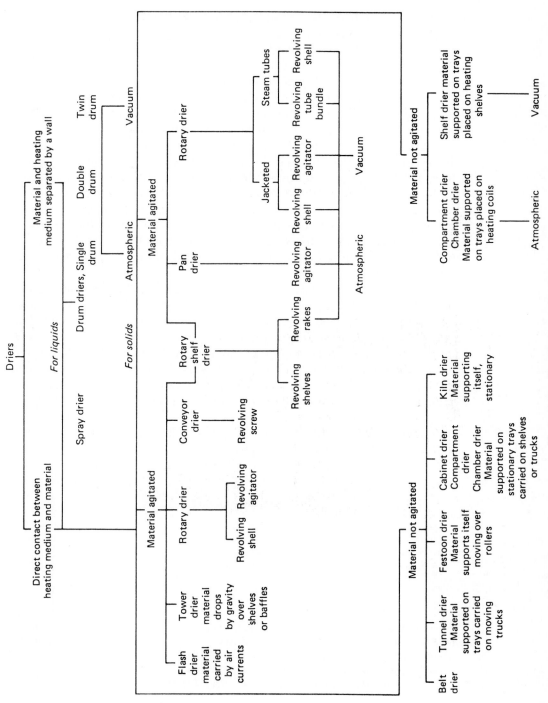

Figure 3.10 Classification of Driers.

Drying a thin slurry is essentially one step: the vaporization of the carrying water (or other liquid), and air in motion (or other gas), in order to carry away the vapor and avoid reaching the saturation point at which vaporization would stop. The task of the first agent is lightened by the application of a vacuum which permits an equivalent amount of vapor to develop at a lower temperature. A means of drying that requires no equipment, namely air drying, may be combined with draining. As draining takes place, there is evaporation to the air, causing the mass to dry over a period of time.

Driers may be divided into two main groups. The first group comprises all the driers in which the material is in direct contact with the heating medium; the second one includes the driers in which material is separated from the heating medium by a wall. The heating medium for the first group may be hot air or flue gas; for the second group, it may be steam, hot water, or flue gas. Each group may be subdivided into driers for liquids and for solids, and those for solids into driers with and without agitation. In many processes drying is the final operation; for this reason it is not merely a matter of removing moisture.

An entirely satisfactory definition of drying is almost impossible to formulate without overlapping definitions of evaporation or distillation, or leaving some drying methods not covered by the definition. Classification of the various types of drying equipment is equally difficult; the one which has been given is based on the method of transferring heat. Driers of similar design generally can be found in both divisions. This classification does not serve as a guide in selecting a suitable drier for a specific product. An alternate classification can be based on the nature of the material handled. (See Figure 3.11.)

The first class of driers comprises those drying units in which solid materials are dried and the individual particles do not move in relation to

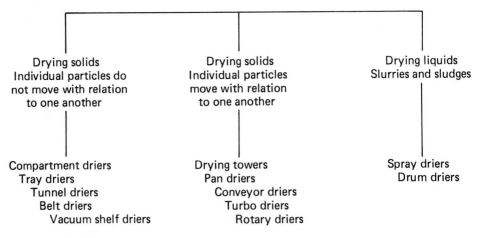

Figure 3.11 Alternate classification of driers based on the nature of the material handled.

one another. These driers are selected for drying materials whose physical shape must be preserved, or which must be dried under carefully controlled conditions for definite time intervals, such as vegetables, fruit, and many other food products; compartment, tray, tunnel, belt, and vacuum shelf driers are included in this class. The drying problem in these driers is largely one of air conditioning, temperature and humidity control, and air circulation.

In the second class of driers also solid materials are handled, but individual particles do move in relation to one another; that is, the material is continuously agitated. The problem in these driers is largely one of agitation and material handling. Included in this class are pan, conveyor, turbo, and rotary driers of various types with revolving shells, revolving agitators, or tube bundles.

The third class of driers is for drying liquids, slurries, and sludges sufficiently fluid to be pumped; drum driers, spray driers, and some others are included. The most important problem in this class is that of the dispersion or spreading of the liquid.

Sludges and pastes which cannot be pumped can be treated as liquids by adding water, or as solids by mixing them with a sufficient quantity of the dry product.

The main cost in drying operations is for heat. In all driers with direct contact between heating medium and material there is a considerable heat loss, namely, all the heat that was required to raise the gases from room temperature to the temperature at which they are discharged from the drier. The efforts to reduce this loss have led to recirculation and to reheating of the exhaust gas. When flue gases can be used, this loss is smaller, and compared to steam, the cost of raising the latter is eliminated, so that direct-heat driers are economical.

Spray Driers

The spray drier consists of a closed chamber in which a liquid carrying dissolved solids or a thin slurry carrying solids in suspension is atomized. A current of hot air meets the droplets and travels with them or past them, simultaneously vaporizing the water, so that when the bottom, or the discharge point is reached, only a dry solid remains in fine suspension in an air laden with water vapor. The flow of air may be parallel to the flow of the droplets and downward, which permits complete drying of the material by the time it reaches the bottom. The dried particles are so light that they follow the air currents. The dried product accumulates in the bottom of an inverted cone, for example, and is pulled away by an exhauster, if not by the escaping air itself which delivers it to a collector and bin. The air with all its moisture travels the same path and is discharged from the collector into the atmosphere. When the flow of air is counter to that of the material, it is essential that the air be made to traverse a curtain of atomized liquid in a wet scrubber, which then retains the entrained particles of dry matter.

The spray drier, then, consists of a drying chamber, an atomizing device, means for moving and heating the air (or other gases), and for recovering the product. (See Figure 3.12.) The drying chamber may be 10 to 20 feet in diameter, and 15 to 30 feet in height; it may be in the form of a cone or of an upright cylinder. The air is heated by passing it over a nest of finned or gilled tubes in which steam circulates. More generally, however, flue gases are used as a source of heat, with a similar arrangement of tubes, except that it is now flue gases which pass inside them (indirect application). It has been found feasible to go a step further and mix air with flue gases from an oil or gas burner (direct application). Steam tubes give an air temperature of 300°F or so; flue gases, indirectly or directly applied, give 500°F and higher, not infrequently as high as 200°F; the temperature is regulated by the volume of cold air admitted. It is desirable to have a high inlet temperature, because the amount of the applied heat transferred to the wet material is then greater, as already suggested under heat economy. For example, let 60°F air be heated by tubes in which 150-pound steam circulates; the steam has a temperature of 366°F. The air will reach 320°F, let us say. If the outlet temperature is 175°F, in accord with general practice, the useful heat is 56 percent of the total heat. With flue gases, directly or indirectly applied, 500°F air or air-flue gas mixture may be obtained. If the outlet temperature is the same as before, the heat

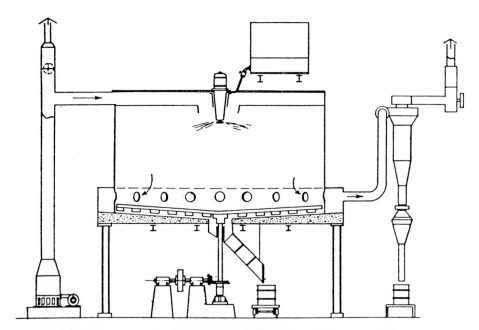

Figure 3.12 A spray-drier assembly, including burner with air inlets for the flue gas-air mixture, the drying chamber with atomizer and feed tank, rake for sweeping the product into the discharge chute, dust collector, and exhauster. The spray chamber is circular and 20 feet in diameter. The flue gas-air mixture leaves the chamber by the annular enlargement at its base.

applied usefully is 74 percent of that put into the air. With 750°F gases, the corresponding figure is 83 percent. The direct use of flue gases is not only more economical, but it has the further merit of reducing the danger of dust explosions because of the lower oxygen content of the gases.

The temperature which may be used depends on the sensitivity of the material. If it scorches easily, a more moderate temperature will have to be selected. In driers in which hot air travels parallel to the droplets, the hottest air comes in contact only with the most wet material, which is protected by its water. As evaporation proceeds, the temperature of the air falls; and it must also be remembered that the droplet is always cooler than the air surrounding it. As the droplet turns to a semisolid, then to a solid, the temperature of the air has dropped so far that there is little danger of scorching.

The total time the droplet spends in the system, from liquid to collected powder, is between 15 and 30 seconds. The period of active evaporation is probably 4 seconds. The spray drier, often called the instantaneous drier, is particularly well adapted to drying heat-sensitive substances.

Parallel and downward flow of air direction was mentioned. An increase in the velocity due to an effort to increase output causes the formation of eddy currents and, along the outer vertical walls, a return current which results in coatings on the walls. Parallel and downward flow requires larger chambers. In order to avoid eddy currents, the air current is made tangential, skirting the wall first and traveling inward toward the center bottom exit, for example, or toward peripheral outlets near the base. In the inverted, cone-shaped chamber, the air may enter tangentially, continue downward along the walls, which gradually approach each other, then turn about at the apex of the cone and travel upward to the exit flue in the center top.

The reason the air is discharged at what may appear to be a needlessly high temperature and incompletely saturated with water vapor is that reabsorption of moisture must be avoided. It is essential that the air be kept well above its dew point; otherwise it might water on the product. The heated air (or air-flue gas mixture) must be fed to the drier in sufficiently large volumes to prevent its reaching saturation, even at the discharge temperature. The high outlet temperature is necessary for the parallel flow of product and air; the residual moisture in the dry product depends to some extent on the outlet temperature. To make certain that this moisture will not be too high, the outlet temperature is not allowed to drop below a safe point. The resulting poor heat economy is remedied to some extent by installing economizers in which the incoming air is heated by the outgoing gases or by setting up scrubbers in which the outgoing gases meet the incoming liquid. In the scrubber, any fine particles carried by the air are deposited, while at the same time some of the water is evaporated from the liquid. Another way to improve the economy is to recirculate and reheat part of the exhaust gas.

Dropping from 300°F to 175°F, one pound of air will give up 29.8 Btu and evaporate 0.030 pound of water; from 500°F to 175°F, it will evaporate 0.078 pound and from 750°F to 175°F, 0.137 pound of water. Inasmuch as air which is saturated at 175°F carries over 0.5 pound of water in vapor form for each pound of dry air, there will be little danger of condensation at this temperature. It may be well to emphasize that air-flue gas mixtures contain considerable amounts of moisture formed as a product of the combustion, a circumstance which must be given consideration.

There are three kinds of devices for atomizing the liquid: (1) the pressure nozzle of special construction designed to impart a rotary motion to the liquid as it leaves the tip of the nozzle to which the spray liquor is fed under pressures varying from 100 to 1,000 pounds per square inch, and which delivers a cone-shaped rain of droplets; (2) the two-fluid type nozzle in which the liquor meets the atomizing steam (or air) under 100 pounds pressure (of the steam or air); (3) the centrifugal atomizer, a disk rotating as fast as 10,000 rpm. The last two are less easily clogged by dirt.

The product may be removed from the bottom of the chamber by a separate fan, or by the outgoing gases, whose velocity in the smaller flue provided is high enough to convey the solid to a cyclone separator, followed by a bag filter or scrubber or electrostatic precipitator. The product may also be removed by scraper arms which deliver it to a mechanical conveyor.

The heat input to the drier required for the evaporation of one pound of water from the spray liquid varies from 1,800 Btu to as high as 3,500 Btu. As to capacity, the spray drier has been built in sizes which will allow the production of 40 tons or more of dry product per day. Spray driers do more than dry, they give the product a powdery quality; certain materials on spray drying form hollow spheres. This indicates again that drying is a unit process. The physical quality of the product is likely to determine whether spray drying is to be adopted, regardless of heat consumption. It should be remarked that the shape of the product is greatly influenced by the type of nozzle and, with certain materials at least, by the speed of the atomizer.

The advantages of spray drying are continuous operation, the short period of contact, which is well indicated by speaking of the spray drier as the instantaneous drier, and the safe handling of heat-sensitive materials.

The spray drier has difficulties, among which may be the tendency for the material to coat the walls of the drying chamber. Two remedies are: one using two currents of drying air, both tangential but opposite to each other, and one outside the other. A greater efficiency in air application is made possible: The humidity of the gas leaving may be made greater than 50 percent compared to about 5 percent saturation obtained with a desiccating apparatus employing a single vortex. Also a reduction in the size of the spray chamber is possible. (See Figure 3.13.)

Materials of Construction

The spray chamber, which is the essential part of the spray drier, may be built of any plate metal such as galvanized iron, stainless steel, Monel metal, and others. For a well-designed unit with a 10-foot diameter spray chamber,

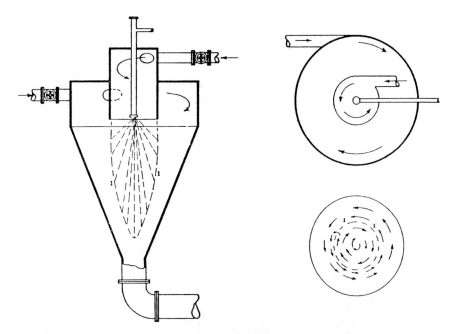

Figure 3.13 A cross-sectional view of the double-vortex spray chamber. Upper right, a top view showing the opposite directions of tangential flows; lower right, the supposed path of a particle.

with air of 300°F inlet temperature, and liquor containing 33 percent solids preheated, there would be a production of 500 pounds of dry product (2 percent residual moisture) per hour, or 6 tons a day. In the same unit, with the same liquor, but with an inlet temperature of 1,000°F, the production would be 2,500 pounds per hour, depending on the type of auxiliary equipment.

Spray Chilling

If underdrying is included as the transformation of liquid water to solid water to water fixed as water of crystallization, then spray chilling too is a form of drying. Spray chilling is the atomization of a solution of crystallizing strength which is liquid while hot, but has only as much water as is needed to form the solid crystal when cooled. The hot solution is atomized in a chamber, which is essentially a spray drier, and meets a current of cold air, this time with parallel flow. The finely divided crystal duct collects on the bottom and in a bag filter or similar auxiliary device.

Flash Driers

The flash drier represents the application of the principle of the spray drier to materials which are solid or semisolid in the wet state. The wet material is dried while suspended in finely divided form in a current of heated air. This is accomplished by dropping the wet material into a high-temperature air stream which carries it to a hammer mill or high-speed agitator where the exposed surface is increased. The fine particles leave the mill through a duct small enough in area to maintain the carrying velocities and reach a cyclone separator. The flash drier is thus another example of parallel-flow operation.

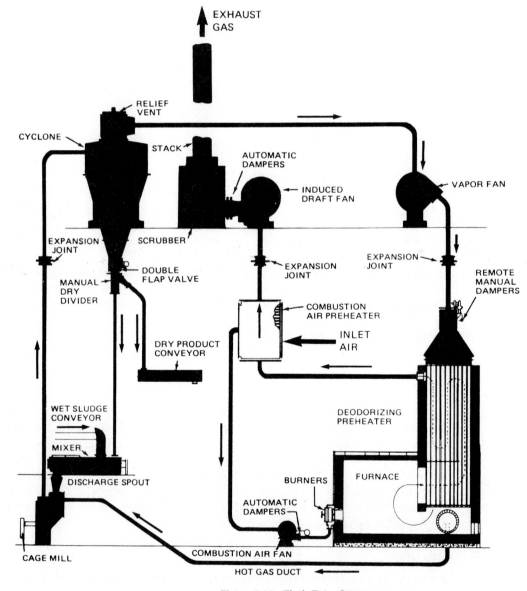

Figure 3.14   Flash Drier System.

(See Figure 3.14.) The particle takes 6 to 8 seconds to pass from the point of entry into the air stream to the collector. The air temperature drops from 1,200°F to 600°F, for example, in 2 seconds, or from 1,200°F to 350°F in 4 seconds. The time for drying is essentially 2 to 4 seconds; hence, the term *flash*. The material itself is not raised in temperature by more than perhaps 100°F; thus, materials which would burn at the temperature of the heating agents pass through unharmed. An initial moisture content of 80 percent may be reduced to 5 percent or 6 percent in the dried product.

In the Raymond flash drier, the wet solids are caught in a flow of high-velocity flue gases (3,000 to 4,000 feet per minute), travel through a special squirrel-cage disintegrator, where the many new surfaces are constantly exposed to the hot gases, and thence to a cyclone separator. The dried solid is collected, and in many installations is mixed with the wet incoming solid in the ratio of 2 to 1.

Drum Driers

In drum drying, a liquid containing dissolved solids or a slurry carrying suspended solids is spread on the surface of a large drum lying on its side and heated internally. In the simplest arrangement, the drum dips in a pan underneath and there receives its coat. As the drum revolves, the liquid is gradually vaporized, so that after seven-eighths of a revolution, a dried deposit can be scraped off (sometimes loosening is sufficient) by a flexible, adjustable knife. The rotation of the drum is made slow enough so that all of the liquid portions are evaporated as the residual solid approaches the knife. The solid collects on an apron in front of the knife and rolls to a container or to a screw conveyor (see Figure 3.15).

The drum is rotated continuously by a gear driven by a pinion which receives its motion through a belt, a chain, or through a reduction gear from a direct-connected motor. The speed of the drum may be regulated by a

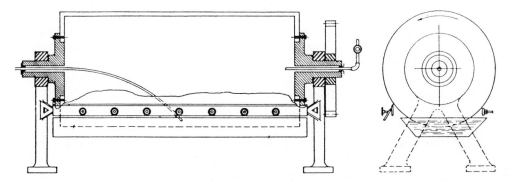

Figure 3.15 An atmospheric single-drum drier with dip feed. Besides the knife, a spreader is shown which regularizes the coating. This gives some of the details of construction which insure sturdiness; support for pan not shown.

variable-speed drive to adapt the speed to any slight variation in the liquid. If the material is dry quite a distance before the knife is reached, the speed should be increased; if the material is too wet at the knife, the speed must be decreased.

The knife may be held just against the surface, or it may be forced against it by turning the adjusting wheels. In recent installations, the knife supports may be turned through part of a circle so that the angle of the blade of the knife relative to the drum surface may be selected for the greatest shearing effect.

The drum drier is heated by steam which enters through the trunnion. The condensate is discharged by means of a scoop or syphon through the second trunnion. The rotation of the drum varies with its duty, but in general it will lie between 4 and 10 rpm. Occasionally, it is slower, and sometimes faster; the limits of from 1 to 20 rpm will cover all installations.

The drum for the single atmospheric drum drier is made of a single casting. The face is turned true on a lathe and is then rolled and polished. The materials are cast iron, bronze, or are chrome plated. For dimensions, a drum 5 feet in diameter and 12 feet long would be a large-sized one. The smallest commercial size is generally a 24-inch diameter drum, 24 inches to 36 inches long. The heat from the condensing steam passes through the condensate film, through the metal of the drum, and then through the coating on the drum. The maximum rate of evaporation for a dilute solution, which gives up its water more readily than a concentrated one, is as high as 18.5 pounds of water per hour per square foot of drum surface. This indicates that a high rate of evaporation may be obtained. As the purpose of drum drying is rather to produce a quantity of dry material, the true measure of efficiency is the number of pounds of finished product per unit heating surface. The rate of evaporation is determined by the concentration at which the material is fed to the drier.

Capacity of a drum drier depends on its dimensions, the speed of rotation, and on the initial concentration of the liquor or slurry. It will also depend on the residual moisture allowable in the product, on the heat resistance of the liquor film, on the steam pressure, and on the adhesion of the coat to the surface. For the last factor, the better the adhesion, the heavier the coating and, therefore, the greater the amount dried in unit time for a given drier. Some of the factors are interrelated; also the result is not always the one expected. For instance, with increased speed, which would lead to expectation of a higher output, the coating is generally thinner, so that capacity does not increase in proportion. Increased steam pressure increases the available temperature difference and in most cases the adherence also; thus, the capacity increases more than in proportion to the increased temperature difference. Greater concentration increases the thickness of the coating, but for many materials there is a critical point beyond which the adhesion or the uniformity of the coating is adversely affected. Residual moisture increases with higher speed, lower steam temperature, greater thickness of the coating, and higher concentration. As to the dimensions of the drum, the capacity is practically proportional to the surface area. It is slightly favored in the larger sizes because of better mechanical control of operating conditions.

Capacity for various products varies greatly, as adherence and liquid film resistance (to heat) vary over an extremely wide range. The amount of dry product may be from one to six pounds per square foot of surface per hour. The method of application of the liquid to the drum surface, briefly described later, is very important in its effect on adhesion. The size of the drum required for a given duty is made after trials on a small drum, by determining the output per square foot of surface. A drum is then chosen whose dimensions afford the number of square feet necessary. The time interval from dip to knife on the trial drum becomes the time interval on the larger drum. Peripheral speed has no relation to the results.

The heat economy of drum driers is favorable because the heat of condensing steam is transferred directly to the material. A sufficient quantity of heat has to be furnished to compensate for radiation loss. This may be calculated on the basis of a pound of dry material produced and not on the basis of a pound of water evaporated, for when computed on that basis, it will be greater for concentrated than for dilute solutions—a misleading index.

It may be desirable to avoid dipping the heated drum into the liquor to be dried in order to prevent boiling of the liquid or undue concentration. One

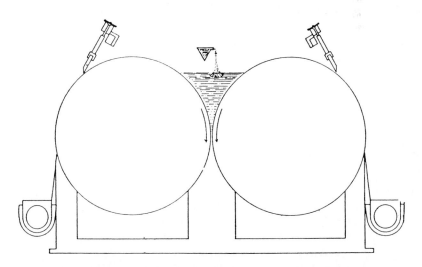

Figure 3.16 An atmospheric double-drum drier with center feed. The drums rotate at the rate of 4 rpm; the drum tops rotate toward each other. The dry material collects in the conveyors which run along the face of the drums. Driving gear and vapor hood not shown. End boards form the two end walls which confine the liquid or slurry.

way is to use a type of feed in which the liquor is splashed onto the drum from below by splash rolls. Another way is to use the double-drum drier. It consists of two drums set closely together, revolving so that the drum tops move toward each other, and with only enough clearance to pass. The material is fed into the trough between two drum tops and two end boards which confine the liquor. Thin pastes may be applied in this way; thicker pastes also may be used with the aid of a special feeding trough with an agitator which delivers a definite amount to the V-space between the drums. The location of the discharging knife is shown in Figure 3.16.

The amount of product from double-drum or twin-drum driers varies from 1 to 10 pounds per square foot of surface per hour, depending almost entirely on the properties and the concentration of the material. (See Figure 3.17.)

Steam pressure is one of the factors affecting capacity. As higher steam pressure means accelerated evaporation, and as there is a limit to the tensile strength of the castings, it may be possible to choose a double-drum or a twin-drum drier of smaller surface and still obtain the same production as on a single-drum drier of larger surface. Thus, a single drum 5 feet in diameter and 144 inches wide has a surface of 27,000 square inches; with a double-drum drier with drums 42 inches in diameter and 100 inches wide, each drum has a surface of 13,400 square inches. The single drum receives 50-pound steam, which is the maximum pressure it can safely take. The double-drum may take 100-pound steam pressure safely. The capacity in dried material will be the same.

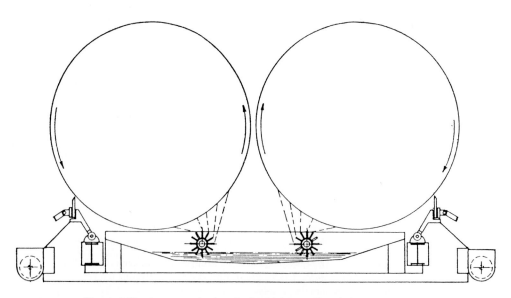

Figure 3.17 An atmospheric twin-drum drier with splash feed. The rotation of the drums is opposite to that in the double-drum drier.

The drum metal should be of minimum thickness and maximum conductivity compatible with safety. Unfortunately, most metals of high strength, such as stainless steel, or high conductivity, such as copper, are comparatively soft; they are not used except in a few installations. On the other hand, cast iron has been so greatly improved in strength and uniformity that it may now be built in lighter sections. The strength of steel can now be made available by chromium plating its surface, for in that way the required hardness is provided, together with resistance to corrosion. The drum drier may be equipped with a vapor hood and exhaust fan, pulling out the water vapor with fresh air.

Rotary Driers

The rotary drier is a cylinder, slightly inclined to the horizontal, which may be rotated as in the revolving rotary drier, or the shell may be stationary and an agitator may revolve slowly. In either case, the wet material is fed in at the upper end, and the rotation (or agitation) advances the material progressively to the lower end where it is discharged. In the direct-heat, revolving rotary drier, warm air or a mixture of flue gases and air travels through the cylinder. The rate of feed, the speed of rotation or agitation, the volume of heated air or gases, and their temperature are so regulated that the solid is dried just before discharge.

The shell fits loosely into a stationary housing at each end. The material is brought to a chute which runs through the housing; the latter also carries the exhaust pipe (see Figure 3.18). The revolving shell runs on two circular

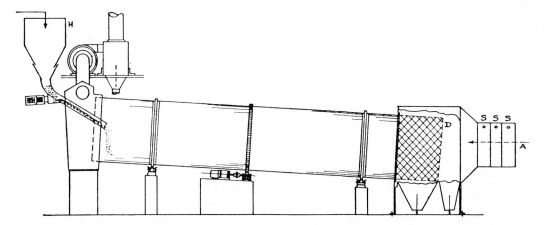

Figure 3.18 A direct-heat rotary drier, 9 feet in diameter and 45 feet long. The shell runs on two circular rails, carried by two sets of four idlers. The girth gear and its driving motor and connection are at the center of the shell. The wet material enters hopper H and the vibrating feeder; the dry material enters the delumper D, which retains the lumps, but permits the fines to drop to hopper below, which feeds a bucket elevator. The combustion gases enter at A, with secondary air regulated by inlet baffles S; the gases travel through the shell, and leave through the exhauster at upper left.

tracks and is turned by a gear which meshes with a driven pinion. The incineration is 1 in 16 for high capacities and 1 in 30 for low ones.

As the shell revolves, the solid is carried upward one-fourth of the circumference; it then rolls back to a lower level, exposing fresh surfaces to the action of the heat as it does so. This simplest of revolving rotary driers serves well enough when the fuel is cheap. The efficiency is greatly improved by placing longitudinal shelves 3 or 4 inches wide on the inside of the cylinder (itself, let us say, 5 feet in diameter). The longitudinal shelves are called lifting flights. The shelves carry part of the solid halfway around the circumference and drop it through the whole of a diameter in the central part of the cylinder where the air is hottest and least laden with moisture. By bending the edge of the shelves slightly inward, some of the material is delivered only in the third quarter of the circle, producing a nearly uniform fall of the material throughout the cross section of the cylinder. The heated air streams through a rain of particles. This is the most common form of revolving rotary cylinder and has a great capacity, is simple to operate, and is continuous.

For fine materials, which are powders when dry and which must be dried to a low-moisture content, the air speed must not exceed 2 or 3 feet per second. These are the velocities in chemical driers. For certain other purposes, such as the drying of lump coal, the velocity may be 12 or 15 feet per second.

The mass of material in the drier is 30 percent or 40 percent of the volume of the shell; the air has a narrower channel than the full diameter of the shell, and for the same volume of air the velocity is higher. Velocities are commonly computed on the basis of the whole shell space.

Division of the falling particles, which permits more intimate contact between the hot air or hot gases and the material, is further increased in the cellular type of rotary drier whose lifting flights have become cells filling the interior of the shell. As the shell rotates, the material flows from one cell to another, thoroughly exposing the wet surfaces. The cellular drier produces less dust in the exit gases than does the rotary with lifting flights; it is more difficult to repair and higher in initial cost. A schematic for a rotary drier is shown in Figure 3.19.

The direction of air travel can be given in three examples of revolving rotary driers and is generally countercurrent to the flow of the material; the hottest air enters at the lower end, and leaves where the wettest material enters. However, the hot air travels concurrently, entering at the upper or feed end and leaving at the lower or discharge end. When the material is sensitive to heat, this is obligatory, although the efficiency of heat utilization is somewhat lower. Governing efficiency, however, is the temperature drop of the air or gases, so that here, as in the spray drier, the higher the inlet temperature of the air or gases, the greater the efficiency. Thus, steam-heated air entering the drier at 200°F, for example, and leaving at 120°F, using 60° air for the heater, would usefully apply 57 percent of the heat. Based on the steam used in the heater, it is nearer 40 percent. A high-temperature, direct-

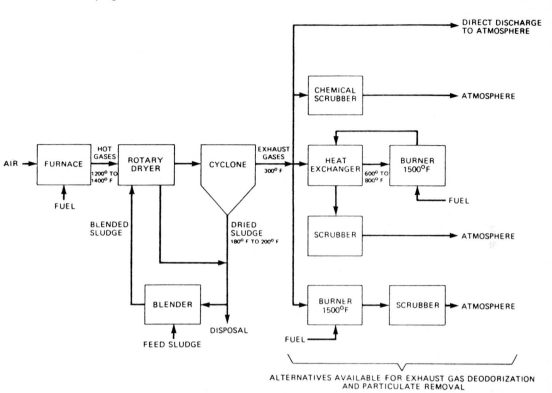

Figure 3.19 Schematic for a rotary drier.

heat, revolving rotary drier with an inlet temperature of 700°F, for example, and direct application of the fuel heat may usefully apply 80 percent or 85 percent of the heat.

Belt Driers

A belt drier consists of a set of hinged shelves or mesh wire between two endless chains. As the belt travels on large sprocket wheels, its path lies within a chamber in which warm gases circulate. The belt is long enough to keep the product within the heated chamber for a prescribed time after which time, it is carried to discharge. The purpose of the belt drier is to allow placing material in and taking it out of the drier with a minimum of labor. A belt-conveyor drier is a belt drier with the belt made up of stout links. The belt travels and the product reaches a long compartment or tunnel in which warm air is circulated. As the belt moves, the dried cake or sludge approaches the point of discharge, and the time is so computed that it arrives there dry.

Conveyor-type driers consist of a belt made of woven wire of suitable mesh, on which the material is spread; the drying air passes through the bed of material and through the meshes. The belt travels on a chain or is simply stretched over two large pulleys; the whole assembly is generally enclosed in a cabinet. The belts are made of steel wire or of any corrosion-resistant metal obtainable in wire form; the manufacturers of this special belting have also developed a belt with side guards woven into the main belt to confine the material. In order to save space, instead of one very long belt, a series of belts may be arranged one on top of another in a multipass arrangement.

With this type of drier the material may be subjected to stages of varying heat and humidity to accomplish continuously what would otherwise take two or three steps. As an example, a material which tends to case harden may be subjected to a zone of high humidity and high heat to draw moisture out of the center; then the humidity may be lowered in the next zone to dry it thoroughly. These two zones can be supplemented by a cooling zone to prepare the material for packaging. The zones are separated by light-gauge dividing walls. This type of drier lends itself very nicely to recirculation of part of the air after it has been reheated.

Indirect Drying

Indirect rotary dryers have not been used in the United States for drying sludge. Vertical thin-film dryers are used in Europe. Two LUWA double-wall driers installed in France operate on 140 psi (966 kN/m^2) steam at a temperature of about 355°F (180°C). The evaporaters are vertical, with top inlet and bottom outlet. Steam generated from refuse incineration is forced into the drier and heats a "jacket" surrounding the incoming dewatered sludge. The sludge is spread over the inner cylindrical surface of the dryer by a rotor carrying self-adjusting vanes, at a top speed of about 25 feet per second (7.6 m/sec). The water vapor travels upward, counter to the sludge flow, and is blown into the incinerator, where it is deodorized. The dried sludge falls onto a conveyor belt and is incinerated with the refuse.

Direct-Indirect Rotary Driers

The direct-indirect rotary drier is similar to indirect driers employing hot air or gases as the heating medium. In direct-indirect drying, however, the heating medium is recirculated to flow in direct contact with the drying sludge in addition to heating the metal drying surface.

Another type of indirect sludge drier is the jacketed and/or hollow-flight drier and conveyor. A schematic of a jacketed hollow-flight drier is presented in Figure 3.20. These units can perform the dual function of heat transfer and solids conveying in one piece of equipment—generally a horizontal, semi-circular trough with a jacket or coil to provide heat. This equipment has one or more agitation devices (for example, screw, flight, disc, paddle) rotating on the axis through the center of the trough. A significant degree of agitation is necessary to maintain reasonable heat transfer. Simple screw conveyors are notably poor in this regard because increasing the speed reduces the

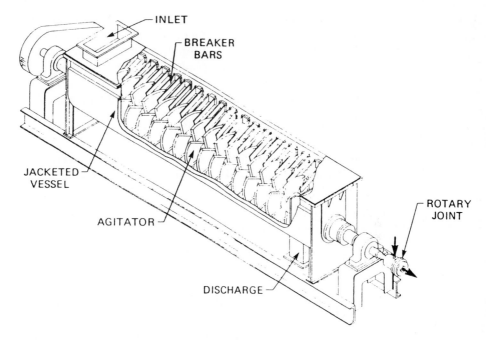

Figure 3.20   Jacketed hollow-flight drier.

residence time in the drier by moving the sludge rapidly through the system. Heat-transfer coefficients for this type of equipment range from 15 to 75 Btu per hour per square foot per °F (18.6 to 93 cal/sq cm/°C), depending on moisture content and degree of agitation.

HIGH-TEMPERATURE PROCESSES

High-temperature processes have been used for combustion of municipal wastewater solids since the early 1900s. Popularity of these processes has fluctuated greatly since their adoption from industrial combustion. In the past, combustion of wastewater solids was both practical and inexpensive. Solids were easily dewatered and the fuel required for combustion was cheap and plentiful. In addition, air-emission standards were virtually nonexistent. Today, wastewater solids are more complex and include sludges from secondary and advanced waste treatment processes. These sludges are more difficult to dewater and thereby increase fuel requirements for combustion. Due to environmental concerns with air quality and the energy crisis, the use of high-temperature processes for combustion of municipal solids is being scrutinized.

More efficient solids dewatering processes and advances in combustion technology have renewed an interest in the use of high-temperature processes for specific applications. High-temperature processes should be considered where available land is scarce, stringent requirements for land disposal exist, destruction of toxic materials is required, or the potential exists for recovery of energy, either with wastewater solids alone or combined with municipal refuse.

High-temperature processes have potential advantages over other methods which include:

- Maximum volume reduction. Reduces volume and weight of wet sludge cake by approximately 95 percent, thereby reducing disposal requirements.

- Detoxification. Destroys or reduces toxics that may otherwise create adverse environmental impacts.

- Energy recovery. Potentially recovers energy through the combustion of waste products, thereby reducing the overall expenditure of energy.

Disadvantages of high-temperature processes include:

- Cost. Both capital and operation and maintenance costs, including costs for supplemental fuel, are generally higher than for other disposal alternatives.

- Operating problems. High-temperature operations create high maintenance requirements and can reduce equipment reliability.

- Staffings. Highly skilled and experienced operators are required for high-temperature processes. Municipal salaries and operator status may have to be raised in many locations to attract the proper personnel.

- Environmental impacts. Discharges to atmosphere (particulates and other toxic or noxious emissions), surface waters (scrubbing water), and land (furnace residues) may require extensive treatment to assure protection of the environment.

High-Temperature Operations

Combustion is the rapid exothermic oxidation of combustible elements in fuel. Incineration is complete combustion. Classical pyrolysis is the destructive distillation, reduction, or thermal cracking and condensation of organic matter under heat and/or pressure in the absence of oxygen. Partial pyrolysis, or starved-air combustion, is incomplete combustion and occurs when insufficient oxygen is provided to satisfy the combustion requirements. The basic elements of each process are shown on Figure 3.21. Combustion of wastewater solids, a two-step process, involves drying followed by burning.

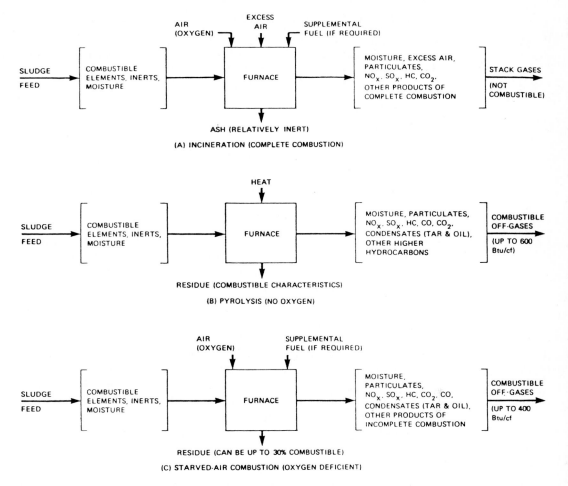

Figure 3.21 Basic Elements of High Temperature Processes.

Sludge Fuel Values

A value commonly used in sludge incineration calculations is 10,000 Btu per pound of combustibles (see Table 3.3). It is important to clearly understand the meaning of combustibles. For combustion processes, solid fuels are analyzed for volatile solids and total combustibles. The difference between the two measurements is the fixed carbon. Volatile solids are determined by heating the fuel in the absence of air. Total combustibles are determined by ignition at 1,336°F (725°C).

The difference in weight loss is the fixed carbon. In the volatile-solids determination used in sanitary engineering, sludge is heated in the presence of air at 1,021°F (550°C). This measurement is higher than the volatile-solids measurement for fuels and includes the fixed carbon. Numerically, it is nearly

TABLE 3.3 CHEMICAL REACTIONS OCCURRING DURING COMBUSTION

Reaction	High Heat Value of Reaction
$C + O_2 \longrightarrow CO_2$	$-14,100$ Btu/lb of C
$C + \frac{1}{2}O_2 \longrightarrow CO$	$-4,000$ Btu/lb of C
$CO + \frac{1}{2}O_2 \longrightarrow CO_2$	$-4,400$ Btu/lb of CO
$H_2 + \frac{1}{2}O_2 \longrightarrow H_2O$	$-61,100$ Btu/lb of H_2
$CH_4 + 2O_2 \longrightarrow CO_2 + 2H_2O$	$-23,900$ Btu/lb of CH_4
$2H_2S + 3O_2 \longrightarrow 2SO_2 + 2H_2O$	$-7,100$ Btu/lb of H_2S
$C + H_2O \text{ (gas)} \longrightarrow CO + H_2$	$+4,700$ Btu/lb of C
Sludge combustibles $\longrightarrow CO_2 + H_2O$	$-10,000$ Btu/lb of combustibles

the same as the combustibles measurement. In the following, if volatile solids is used in the sense of the fuels engineer, it will be followed parenthetically by the designation *fuels usage*. If the term *volatile solids* or *volatiles* is used without designation, it will indicate sanitary engineering usage and will be used synonymously with *combustibles*.

The amount of heat released from a given sludge is a function of the amounts and types of combustible elements present. The primary combustible elements in sludge and in most available supplemental fuels are fixed carbon, hydrogen, and sulfur. Because free sulfur is rarely present in sewage sludge to any significant extent and because sulfur is being limited in fuels, the contributions of sulfur to the combustion reaction can be neglected in calculations without compromising accuracy. Similarly, the oxidation of metals contributes little to the heat balance and can be ignored.

Solids with a high fraction of combustible material (for example, grease and scum) have high fuel values. Those which contain a large fraction of inert materials (for example, grit or chemical precipitates) have low fuel values. Chemical precipitates may also exert appreciable heat demands when undergoing high-temperature decomposition. This further reduces their effective fuel value.

The following are experimental methods from which sludge heating value may be estimated or computed:

- Ultimate analysis—an analysis to determine the amounts of basic feed constituents. These constituents are moisture, oxygen, carbon, hydrogen, sulfur, nitrogen, and ash. In addition, it is typical to determine chloride and other elements that may contribute to air emissions or ash-disposal problems. Once the ultimate analysis has been completed, Dulong's formula can be used to estimate the heating value of the sludge. Dulong's formula is:

$$\text{Btu/lb} = 14,544\,C + 62,208\left(H_2 - \frac{O_2}{8}\right) + 4,050\,S$$

where C, H_2, O_2, and S represent the weight fraction of each element determined by ultimate analysis. This formula does not take into account endothermic chemical reactions that occur with chemically conditioned or physical-chemical sludges. The ultimate analysis is used principally for developing the material balance, from which a heat balance can be made.

- Proximate analysis—a relatively low-cost analysis in which moisture content, volatile combustible matter, fixed carbon, and ash are determined. The fuel value of the sludge is calculated as the weighted average of the fuel values of its individual components.

- Calorimetry—this is a direct method in which heating value is determined experimentally with a bomb calorimeter. Approximately 1 gram of material is burned in a sealed, submerged container. The heat of combustion is determined by noting the temperature rise of the water bath. Several samples must be taken and then composited to obtain a representative 1-gram sample. Several tests should be run, and the results must be interpreted by an experienced analyst. New bomb calorimeters can use samples up to 25 grams and this type of unit should be used where possible.

The preceding tests give approximate fuel values for sludges and allow the designer to proceed with calculations which simulate operations of an incinerator. If a unique sludge will be processed, or unusual operating conditions will be used, pilot testing is advised. Many manufacturers have test furnaces especially suited for pilot testing.

4

Sterilization by Radiation ═══════════

Two types of radiations used in sterilization are ionizing and nonionizing. Generally speaking, the term *radiation sterilization* usually implies the use of ionizing radiation.

IONIZING RADIATION

Ionizing radiations are highly lethal and their use demands the strictest safety precautions. They may be either in the form of particulate high-energy electrons or electromagnetic gamma rays. (See Figure 4.1.) High-energy electrons are cathode rays produced by a high-voltage potential in an evacuated tube. They are accelerated by electrostatic forces and the beam of electrons is directed through a window over a defined target area.

Gamma rays resemble hard (short wavelength) X-rays and are emitted at one or two fixed wavelengths by radioactive elements. The usual source is Cobalt-60 (half-life 5–3 years), a by-product of atomic fission reactions.

Mode of Action

In contrast to heat, radiation injury does not cause denaturation of protein but induces ionization of vital cell components, particularly the deoxyribonucleic acid (DNA) of the nucleus. The extent of ionization of the irradiated molecules depends on the energy level of the radiation used. High-energy

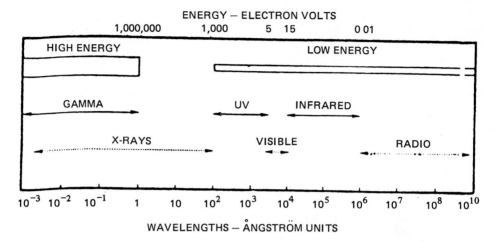

Figure 4.1 Wavelengths and energy levels of various types of electromagnetic radiations.

radiations above 5 Mev may induce radioactivity in the irradiated material and are, therefore, unsuitable for sterilization of medical equipment or foods.

Two theories have been proposed to explain the lethal action of ionizing radiation. The first is the target (direct-action) theory in which it is suggested that ionization of DNA is directly induced. This is supported by the exponential relationship between dose and effect. The diffusion (indirect-action) theory postulates primary ionization of water molecules in the cell, thereby producing free hydrogen and hydroxyl radicals which, in turn, initiate secondary reactions in DNA molecules. Both theories may be reconciled with one another by postulating that the target molecules are surrounded by a water film.

In addition, many other chemical changes may also be induced and these often result in the development of undesirable food flavors, discoloration of glass and fabrics, loss of fiber strength, and the release of chlorine from plastics such as polyvinyl chloride.

Sterilizing Dose

It is usual to measure the sterilizing dose from a Cobalt-60 source in Megarads (Mrad) and that from an electron generator in Mrad or kilowatt-seconds. Table 4.1 lists the range of sterilizing doses for different types of microorganisms. Some of these are based on unpublished findings in these laboratories and are shown as killing curves in Figure 4.2. The doses given in Table 4.1 are comparative rather than definitive, especially in the case of viruses. However, it can be assumed that for all general sterilizations, a dose of 2.5 Mrad is sufficient and has a factor of at least 10^6 for highly resistant bacterial spores such as *Bacillus pumilus*.

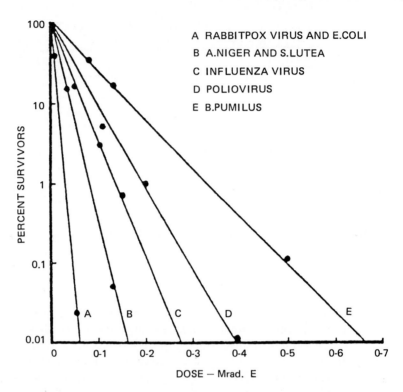

Figure 4.2 Inactivation of different organisms in water by gamma radiation.

Radiation Equipment

Two main types of radiation equipment are available: Cobalt-60 installations and electron accelerators. Both are designed to meet the demands of large-scale industrial production. Their high capital cost precludes any consider-

TABLE 4.1 STERILIZING DOSES FOR IONIZING RADIATION

Group	Organisms	Dose Range (Mrad)
Sensitive	Vegetative bacteria[a] Animal viruses (>75 mμ)	0.05–0.5
Moderately resistant	*Bacillus anthracis* (spores) *Clostridium* sp. (spores) Moulds and yeasts Animal viruses (20–75 mμ)	0.5–2.0
Resistant	*Bacillus pumilus* (spores) *Micrococcus radiodurans* Animal viruses (<20 mμ) Bacterial viruses	2.0–4.0

[a] Including acid-fast bacilli.

TABLE 4.2 CHARACTERISTICS OF RADIATION STERILIZERS

Characteristic	Cobalt-60 Source (Gamma Radiation)	Electron Accelerator (High-Energy Electrons)
Penetration	High—2 feet of water	Low—<1 inch of water
Operation	Continuous	Discontinuous
Sterilization time	Long (*ca.* 48 hours)	Short (seconds)
Materials processed	Wide range—small and large articles	Narrow range—small articles only
Induced radioactivity	Never occurs	Occurs above 5 MeV
Installation	Elaborate, space occupying	Less space required
Safety precautions	Elaborate	Less exacting

ation of their installation in hospitals. A comparison of their characteristics is made in Table 4.2. With both installations the irradiation takes place at room temperature.

A Cobalt-60 installation is housed in a thick-walled concrete chamber. The packages, varying in size from bales of goat hair to cartons of disposable syringes or sutures, are mechanically circulated around the source so that they are irradiated to a calculated dosage of 2.5 Mrad from both sides. The source is lowered into a deep stainless steel tank of filtered water when not in use.

Electron accelerators are of two types: the Van de Graff generator, which has an output of 1–3 Mev, and the microwave liner accelerator, whose output may vary from 3–15 Mev. In contrast to the massive physical dimensions of a Cobalt-60 installation, an electron accelerator is small. A beam of electrons is directed onto a conveyor belt carrying small articles of uniform size, such as single packets of sutures or syringes (Figure 4.3).

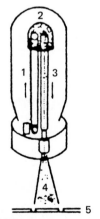

1 MOVING BELT
2 HIGH-VOLTAGE TERMINAL
3 HIGH VACUUM TUBE
4 ELECTRON BEAM
5 PACKAGES ON CONVEYOR BELT

Figure 4.3 Operating principle of the Van de Graff generator.

Application

It is possible that ionizing radiation will eventually become the main method of sterilizing the bulk of commercially packed medical equipment. At present there are only a few industrial installations of Cobalt-60 plants in operation, one of the largest being in Dundenong, Australia. This plant, with a source of 500,000 curies, was built with the primary object of sterilizing anthrax spores in intact bales of goat hair. Table 4.3 lists applications of ionizing radiation.

The advantages of radiation sterilization are:

- The method is highly efficient.
- The temperature rise is negligible.
- Articles may be packaged and sealed before sterilization.

However, there are some limitations in its use, the most important of which are:

- The sterilization time is long if Cobalt-60 is used (48–72 hours).
- Glassware darkens and textile fibers lose their tensile strength.
- Foods acquire an undesirable flavor at sterilizing doses.
- Color and chemical changes occur in certain pharmaceuticals (for example, penicillins).

Items of hospital equipment that have been considered for radiation sterilization are textiles such as blankets and mattresses. Blankets deteriorate after 20 treatments at sterilizing doses but withstand approximately 100 radio-disinfecting doses of 0.5 Mrad.

TABLE 4.3   SOME APPLICATIONS OF IONIZING RADIATION

Material	Type of Radiation	Dose (Mrad)
Disposable rubber gloves	Gamma rays or high-energy electrons	2.5
Dressings	" " " " "	2.5
Needles and syringes	" " " " "	2.5
Plastic equipment	" " " " "	2.5
Sutures		2.5–3.0
Foods	Gamma rays (radio-pasteurization)	0.1–0.5
Goat hair	" "	2.0
Hospital blankets	" " (radio-disinfection)	0.5
Tissue grafts	" "	2.5
Viral vaccines	" "	1.5

NONIONIZING RADIATION

Bright sunlight has some bactericidal activity due to nonionizing ultraviolet (UV) radiation in the sun's spectrum. Ultraviolet radiation is an electromagnetic radiation of low energy, about 5 eV with poor penetrating power. The bactericidal range is 2,400–2,800 A, the optimum being 2,500–2,650 A.

Ultraviolet radiation causes excitation rather than ionization of atoms by raising electrons to a state of higher energy without removing them. When microorganisms are irradiated, it affects vital cell components, particularly the nucleic acids. Photo-reactivation by visible light has been demonstrated, but this phenomenon has no practical significance in UV sterilization.

Ultraviolet Radiation Sterilization

The ultraviolet region of the electromagnetic spectrum that contains the wavelengths from 200 to 3,100 angstroms is commonly referred to as the *abiotic region*. Because of the relatively low energies involved, UV radiation does not readily penetrate most materials. Due to low penetration abilities, the use of ultraviolet light for the purpose of sterilization and disinfection has been limited to the areas of disinfection of water and surfaces.

Purification of Water

Microorganisms withstand considerably more UV in water than in dry air. Commercially available sources of UV offer, at present, limited but important application in industries practicing bacteriological control which require or desire nonchlorinated water as well as in areas where drinking water is or may be unsafe. Commercial UV units are available that allow rates of water from 75 gallons per hour to 20,000 gallons per hour. Due to this method, 100 percent of waterborne bacteria is destroyed. Ultraviolet water purification has been applied to water wells, cisterns, and swimming pools to avoid heavy chlorination, and where biologically pure water is desirable or required.

Ultraviolet radiation is a valuable tool for limited purification of water. Particularly under special conditions, UV has been used successfully to control microfouling communities responsible for slime formation on optical surfaces immersed in sea water. The use of a commercial UV disinfecting system for water has been designed primarily for shipboard use. *E. Coli*, the index organism of water pollution, has been tested in this system. The viruses were inactivated and over 99.9999 percent of the bacteria were killed yielding good potable water.

Ultraviolet radiation is valuable for treating sea water used for shellfish, since chlorine has an adverse effect on the feeding activity of oysters.

Ultraviolet radiation should not be considered as a replacement for efforts to improve individual sources of water. Waterborne organisms surviving the initial exposure to UV in a closed chamber have been found resistant to another treatment with UV.

The application of UV to liquids more absorptive, and possibly more viscous, than water has resulted in the development of special equipment counteracting this condition by allocating adequate time for exposure of a moving thin film of liquid to the radiation. Ultraviolet irradiation of products such as fruit juices, wine, beer, and others has not been achieved on a commercial scale, largely because of limitations imposed by such products. That is, because of suspended-solid particles or chemical instability in clear water, over 50 percent of the energy is lost at a depth of less than 5 cm, and in river water it may be within 1 cm of the surface.

Preparation of Immunizing Antigens

Application of UV to the sterilization of bacterial and viral suspensions for penetration of immunizing antigens was stimulated by the work of the mercury resonance bulb to produce a nonvirulent but immunizing rabies vaccine. Ultraviolet irradiation of viruses resulted in a higher degree of immunity in mice than was shown by phenolized preparations and exhibited no loss of potency after six months at 5°C. Improvement in this procedure was made by exposing a thin film of antigenic material introduced by gravity or under pressure directly to a UV lamp emitting most radiation at 253.7 nm.

Ultraviolet radiation need not destroy the immunizing property of inactivated suspensions of microbial agents, as measured in animals by protection tests. However, the effect of radiation on the reactivity of antigens must be kept in mind. Although UV-inactivated vaccinia virus produced neutralizing antibodies and resistance to challenge in rabbits, it was active in only about half of the human volunteers. Exposure of human, horse, and bovine serum albumins, as well as of tobacco mosaic virus, to UV led to alternation or destruction of their antigenic properties when measured in vitro.

Irradiation of Plasma

Exposure of a thin film of plasma to the high intensity of 253.7-nm radiation from a low-pressure mercury lamp led to encouraging reports of the effectiveness of this method in reducing the hazard of homologous serum hepatitis. The amount of energy used had no observable untoward effect on plasma proteins. No toxic or allergic reactions were encountered that could be attributed to the treated plasma. In cold sterilization using beta-propiolactone and subsequent UV irradiation of all human sera tested for hepatitis suppressed the risks to laboratory personnel using potentially infective reagents.

Use of UV in Industry, Special Situations, and Wastewater

Air is the important vehicle of microorganisms that may become responsible for deterioration and spoilage of many consumable products such as bread, meat, beer, wine, soft drinks, dairy products, and others. Basic principles of protection by UV irradiation are those already discussed and include:

- Reduction of airborne organisms in usually occupied locations.
- Irradiation of the product during processing.
- Destruction of organisms that settle on surfaces and become part of the dust.

Successful application of these principles claimed to raise the quality of apple cider and wine and various other products, such as syrups and soft drinks, and reduced the growth of mold on equipment or on walls as well as on preserved products such as meat in cold storage. An adequate degree of safety has been achieved by treating tonometer prisms with absolute alcohol, and then irradiating the prisms with UV light.

Application of localized high-intensity radiation over production lines is used by the pharmaceutical industry in sterile rooms and hoods, in filling and cupping rooms, in air-duct systems to provide sterile air to working areas, and in any locations or situations where microbial contamination may be a problem. In addition, certain chemicals and plastics were sterilized by UV without producing untoward changes. Certain filter membranes, special plastic-coated instruments, and similar apparatus cannot be subjected to conventional microbial decontamination.

ULTRAVIOLET DISINFECTION—WASTEWATER APPLICATION

Disinfection of wastewater streams has usually been practiced when the receiving waters are used as drinking water sources or for recreational purposes. Activated sludge treatment removes 90 percent to 98 percent of bacteria in the raw wastewater and proper chlorination destroys another 98 percent of the bacteria. The use of chlorine has been questioned because its resulting compounds can be toxic to aquatic life. In addition, toxic chlorinated compounds may be produced during chlorination. Ultraviolet light has been found to be an alternative disinfection method. Ultraviolet light kills or renders bacteria incapable of reproduction by photochemically altering the DNA in the cell. The bactericidal effects of UV have been known for many years. However, unattractive economics have traditionally dictated the use of chlorine.

Principles of Ultraviolet Disinfection

The practical applications of UV are dependent upon the killing action of its radiations on agents such as yeasts, molds, bacteria, rickettsiae, mycoplasma, and viruses. Some of the other effects of UV include increases in the rate of mutation, chromosomal aberration, and changes in cellular viscosity. Ultraviolet radiation affects such vital processes as respiration, excitability, and growth.

Ultraviolet light is a radiation energy. The radiations are usually divided into two main groups: corpuscular and electromagnetic wave radiations. Corpuscular radiations are exemplified by streams of atoms, electrons, and protons. Electromagnetic wave radiations comprise the wide range of radiations from the longest radio waves without a detectable biological effect, through infrared or heat rays, the visible light, to UV, roentgen rays, gamma rays, and secondary cosmic rays. While the speeds of corpuscular radiations differ, electromagnetic rays travel at the same rate of 186,000 miles per second. They do, however, differ in length. The shortest radiation waves are measured in nanometers (nm; 1 nm is equal to 10^{-9} m).

Although radiation travels as a wave, it acts as if it were produced and delivered in discrete amounts of energy called quanta. Quanta of electromagnetic energy, radiant energy, or photons are emitted from radiation sources. The energy of these photons is directly proportional to frequency. The speed c traveled by the radiation waves is determined by the number of oscillations per second (frequency, f) and the wavelength

$$c = f \tag{4.1}$$

The energy quantum E is given by the following:

$$E = hf \tag{4.2}$$

where E is in erg-sec and h is Planck's constant (6.62×10^{-27}).

The energy of a photon thus increases with increasing frequency or with decreasing wavelength. The preceding relation clarifies the relatively inappreciable photochemical and photobiological effects of infrared radiation in contrast to the marked effects of UV. The germicidal effect of UV could result from absorbed radiation and from an indirect effect of toxic compounds produced in the surrounding medium. However, the latter mode of action appears unlikely. Consequently, the availability of large-energy quanta which are poorly absorbed at a particular wavelength will lead to an inefficient reaction. The absorbed quanta must be of a sufficient amount of initiate and maintain a given photo reaction.

Although the UV spectrum includes a range from 15 nm to the rays bordering on the visible spectrum, the region of primary interest is from 220–300 nm. This is often referred to as the abiotic region. Lethal dose values for 50 percent population kills (LO_{50}) of UV for some unicellular organisms have been studied extensively. The so-called action spectra are similar to germicidal wavelengths and those absorbed by nucleic acids. Studies of action spectra of mutagenic effects and retardation of cell division suggest that these conditions are caused by the effect of UV on nucleic acids. Of the components of nucleic acids, sugar phosphates do not significantly absorb UV above 220 nm. Since pyrimidines are much more sensitive to UV than purine bases, the major effects of mutagenic and lethal UV on biological systems are attributed to photochemical transformations of pyrimidine bases. It appears that the sterilizing UV acts on cellular DNA primarily by producing linkages between successive pyrimidines on a DNA strand to form dimers. It has been suggested that they are primarily formed of adjacent thymine residues in the same strand.

Thymine dimers have become specially linked to losses of the transforming ability of bacterial DNA. Cytosine-thymine mixed dimers and cytosine-cytosine dimers have also been identified in DNA from normal and transforming bacterial strains exposed to UV. Also, cellular DNA of irradiated bacteria, protozoa, and mammalian cells have been observed but less frequently than thymine-thymine dimers. These interfere with replication and also lead to the death of the cells. Ultraviolet irradiation of DNA results in the formation of various kinds of photoproducts that may have a disruptive influence on the local integrity of the DNA structure. Damage other than formation of pyridimine dimers may also contribute to the denaturization of irradiated DNA. As yet, there is no clear indication of the other forms this damage takes.

Not much information exists on the photochemistry of ribonucleic acid (RNA). Hydrates and uracil dimers have been observed in irradiated RNA and both products can cause UV inactivation of RNA. Other pyrimidine photoproducts have been detected. However, the nature of their possible inactivation of RNA has not been established.

A linear relationship appears to exist over a large range of values when the percentage of inhibited cells is plotted as a function of the logarithm of the energy dosage (or if the logarithm of the percentage of inhibited cells is plotted as a function of energy dosage). There is a threshold dose above which inhibition increases rapidly. Maximum bacteria effectiveness is manifested at about 265 nm with a sharp decline at about 290–300 nm, and with progressively lower effectiveness through the visible light range.

Method of Production

Ultraviolet radiation can be generated by the passage of an electric discharge through a low-pressure mercury vapor enclosed in a special glass tube (known commercially as a germicide lamp). The principle of all germicidal lamps is the same: electron flow between electrodes through ionized mercury vapor. The arc in a fluorescent lamp operates on the same principle and produces the same type of energy (UV). The difference between the two is that the bulb of the fluorescent lamps is coated with a phosphor compound which converts UV to visible light. The glass used in ordinary fluorescent lamps filters out all germicidal UV. The germicidal lamp is not coated with phosphor and is made of special glass which transmits the UV generated by the mercury in addition to visible light. About 95 percent of the UV is at a wavelength of 253.7 nm. These lamps have a germicidal efficiency of five to ten times that of high-pressure quartz mercury arcs (400–60,000 mm Hg or 0.5–75 atm).

The inactivation of microorganisms as a result of exposure to UV light is expressed as the product of intensity, I, of germicidal energy, and time, t.

$$K = It \qquad\qquad (4.3)$$

where K is the death rate of microorganisms.

It follows that the same exposure can be obtained using either high intensity for a short time or low intensity for a proportionally longer time. Under practical conditions, the exposure time is dictated by the particular situation.

The efficiency of germicidal lamps depends on the surrounding temperature and air movement. Commercial lamps are generally designed to operate most efficiently at ambient temperatures (around 27°C). Passing air currents over them or submerging them in liquid lowers their output. Special lamps which operate efficiently under otherwise undesirable conditions are available commercially.

In addition to the transmission of 253.7 nm, the glass used in low-pressure mercury lamps will transmit a certain amount at 184.9 nm. Energy of this wavelength forms ozone by breaking the bonds of oxygen molecules. The amount of ozone produced is controlled entirely by the transmission of the glass tubes and decreases more rapidly than emission of 253.7 nm with the age of the lamp. The concentration of ozone in a given area is measured in parts of ozone per million parts of air (ppm). The amount of ozone will vary with temperature, humidity, and air movement. The limit of permissible concentration specified by the Council on Physical Medicine of the American Medical Association is 1 part per 10 million in a conventionally occupied environment. Special germicidal lamps are made with glass envelopes allowing transmission of a high concentration of 184.9 nm along with the germicidal 253.7 nm. These are used in industry for the preservation of food. Ozone has a certain degree of antimicrobial action, inadequate to be of practical significance when human beings are subject to it. Ozone is also used in many situations to eliminate or suppress objectionable odors and to impart a clean, fresh smell to the air.

Bacterial tubes depreciate rapidly during the first 100 hours of operation; consequently, commercial tubes are rated initially as though they had already operated for 100 hours. Meters are available to monitor the output of a particular lamp, as well as to measure the intensities of germicidal energy by walls and ceilings.

Commercial germicidal tubes are available for different current ratings. The maximum intensity of a tube is provided at its own surface. Absorption of UV by air is negligible. However, the distance from the source of radiation imposes certain restrictions in the calculation of intensity. An inverse relationship exists between the distance from the source and the intensity of radiation out to about one half of the effective length of the source, and between the square of the distance at greater than this length from the source. Slimline germicidal lamps are instant-starting and utilize a coil filament on each end which operates hot. Lamp life is governed by the electrode life and the frequency of starts. These lamps are used when light UV intensity is desired, as in the treatment of water or air. Cold cathode bactericidal lamps are also instant-starting, but are not affected by the number of starts. Their use life is determined entirely by the transmission of the bulb. They have a longer life than other lamps and perform well at low temperatures. Because of their long life and moderate intensity, they are often used in occupied areas and when frequent starting is desirable. Most bactericidal lamps operate best in still air at room temperature. Ultraviolet light output is measured at an ambient temperature (77°F). Higher or lower temperatures decrease the output of the lamps.

Lamps require periodic cleaning. By wiping them with a cloth dampened in alcohol or ammonia and water, maximum output of UV can be maintained. Oils or waxes should not be used for moistening the wiping cloth. When a lamp drops to about 60 percent of its 100-hour rating, or after it has been used three fourths of its rating time, it should be replaced.

Survival

Some microorganisms have shown a nonlinear dependency at low UV doses. Vegetative cells of *B. megaterium* show a very slight nonlinear dependence followed by an exponential rate. Dormant spores of this organism do, however, show a much greater nonlinear dependence. Figure 4.4 shows this phenomenon. Curves showing large nonlinearity are referred to as sigmoidal. Sigmoidal survival curves suggest that within the context of target theory, several hits are required to inactivate a single organism. Conversely, with the exponential death-rate dependency, only one hit is needed to cause the death of an organism. This phenomenon merely indicates that certain bacteria pos-

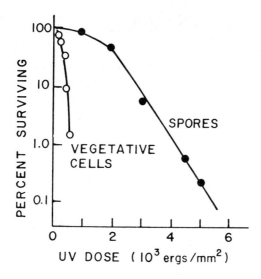

Figure 4.4 Inactivation curves of *B. megaterium* exposed to UV light.

sess the means of repairing cellular damage and are capable of surviving even at high doses of UV radiation.

FACTORS INFLUENCING BACTERIAL SENSITIVITY

Factors which influence bacterial sensitivity to UV light are organism type, dust, and temperature. Bacterial spores are generally more resistant to UV radiation than are vegetative cells. Often, a marked increase in resistance can be observed for some short period. This normally occurs during the germination period. Gram-negative rods are organisms most likely to be killed by UV light; stapllylococci and streptococci requiring about five to ten times, bacterial spores about ten times, and mold spores 50 times as much irradiation for destruction. At low-UV doses, the dose ratio at the same survival for spores and vegetative cells of *B. subtilus* was about 20–40 times *B. megaterium* spores, which are more resistant to UV light than are *B. subtilus* spores.

Dust has been found to affect sensitivities of some microorganisms. *B. subtilus* spores, for example, are more resistant to UV light when in a dust suspension than when exposed to an aerosol.

The sensitivity of bacteria to UV is only slightly influenced by temperatures in the range of 5–37°C. However, at lower temperatures, microorganisms become supersensitive.

PHOTOREACTIVATION

Exposure of UV-irradiated coli to suitable visible light (below 5,100 A) resulted in the recovery of a large portion of the cells from what would otherwise have been death. The effect of reactivating light is photoreactivation.

Photoreactivation is the reversal with near UV or visible light of UV radiation damage to a biological system, or the restoration of UV radiation lesions in a biological system with a wavelength longer than that of the damaging radiation. If N_o is the total number of viable cells before irradiation, N_D (which refers to dark survival) the number of survivors after inactivation, and N_L (which refers to light survival) the number of survivors after photoreactivation, then the fractional light survival is N_L/N_o and:

$$\% \text{ light survival} = 100\, N_L/N_O \qquad (4.4)$$

from which the

$$\% \text{ photoreactivation} = 100 \qquad (4.5)$$

Since not all inactivated cells are capable of being photoreactivated, the percent of photoreactivation cannot, of course, be 100 percent.

There is, however, a wide divergence among the bacteria in their capability to be photoreactivated. Organisms which can be photoreactivated include *S. griseous* and *E. coli* and bacilli.

The diversity in response among bacilli to photoreactivating light is linked with their degree of sensitivity to such light. Spores of photoreactivable strains of vegetative bacilli are not reactivated by light after UV inactivation. Resting spores of *B. cereus* after UV treatment could not be photoreactivated until they were transferred to a nutrient medium on which they could germinate. Sporulating cultures of *B. cereus* completely lose their photoreactivability at the same time as UV resistance increases.

FACTORS AFFECTING TREATMENT EFFICIENCY

The germicidal effect of UV energy depends on its absorption by various organic molecular components that are essential to the cell's physiology. Energy dissipation by excitation causing disruption of unsaturated bonds, particularly of the purine and pyrimidine components of nucleoproteins, appears to produce a progressive, lethal biochemical change. Numerous investigations have been performed to establish the relationship of wavelength, energy, type of organism, and other factors to the lethal effect of UV irradiation on microorganisms. For most species, the bactericidal effect as a function of wavelength is greatest about 2,500–2,600 A, with an abrupt decrease at 2,900–3,000 A.

At any particular wavelength, there seems to be a threshold energy below which the inhibition is negligible and above which the percentage of kill or inhibition rises rapidly at first and then more slowly. Plotting the log or percent of survival cells as a function of energy dose, or percentage of kill as a function of the log of energy dose, approximates a straight line over a wide range of values.

Water Treatment Practice

Until the introduction of more efficient sources of germicidal UV energy at 2,537 A, the possibilities of its practical application to disinfection of water were not realized. With the advent of low-pressure mercury arc "cold" lamps, emitting 85 percent or more of their energy in this region of the spectrum, investigations of radiation requirements and design and testing of equipment for this purpose were carried out by many workers.

There appears to be little question of the germicidal efficiency of UV energy if sufficient dosages reach the organism. Theoretically, water can be disinfected to any degree required. The exponential relation which describes this phenomena is:

$$P/P_o = e^{-Et/Q} \tag{4.6}$$

where

    P = the average number surviving
    P_o = the original number before irradiation
    E = the intensity of germicidal energy
    t = the time of exposure in minutes
    Q = the exposure (Et), termed a unit lethal exposure

Q is found to be approximately 40 microwatt-minutes per cm². For practical purposes, Equation 4.4 appears to represent the relationship adequately, and it has been the basis for the design of a number of sterilizing units with safety factors of two or more.

Factors that affect the penetration of UV energy through water and, hence, the effective destruction of organisms include water turbidity, the presence of iron salts, and organics. Absorption coefficients at 2,537 A are affected mainly by calcium, manganese, sodium, and aluminum ions. The filtering of water samples through coarse to fine-fitted glass filters was found to increase the transmission percentage in many cases. Line voltage fluctuations and ambient temperatures are known to reduce lamp intensity. A drop in voltage from 110 to 100 v can lower the intensity by as much as 22 percent. Temperatures of 0–10°C reduce intensity to 26 percent of that at 20°C with quartz enclosures for the lamps and 42 percent to 78 percent of that at 20°C without quartz enclosures.

Standards have established UV as an acceptable disinfectant for water provided the following criteria are met:

- The UV unit should produce light with a peak intensity of 2,535 A units.
- Lamps should be jacketed in quartz or vapor so a temperature near to 105°F is maintained at the lamp wall. The unit must be designed to provide a means for easy cleaning of the quartz lamp jackets.
- The UV dosage must be no lower than 16,000 MWS/cm².
- Maximum distance between lamps and the chamber should not exceed 3 inches.
- Flow rate must be controlled and not exceed the rated capacity of purifier.
- An accurately calibrated UV intensity meter, properly filtered to restrict its sensitivity to the disinfection spectrum, must be installed in the wall of the disinfection chamber at the point of greatest water depth from the tube or tubes.
- An automatic alarm must be installed to warn of any drop in dosage.
- The material of construction should not impart toxic materials into the water, either as a result of the material or its reaction to UV (plastics would be unsuitable).

At first glance these seemed to be clear, concise, and workable standards. However, in the following years of application, the areas of inadequacy became evident. The standard calling for an UV dosage of 16,000 MWS/cm^2 was the primary source of confusion.

The UV dosage delivered to the water is a function of the intensity of the energy within the purification chamber at the maximum flow rate. Since UV disinfection does not change the water physically or chemically, it is not practical to test the water or alter treatment to measure dosage. Therefore, there is no way that a prospective purchaser of UV purifiers can be certain that the design selected delivers the required dosage. There are currently three basic-size UV lamps which conform with the output requirement of 4.85 UVW/ft^2 at 2 inches. They have effective arc lengths of 30, 58, and 60 inches. The 60-inch lamp will treat 12 gpm.

Table 4.4 gives the required number of lamps and the volume of the purification chamber for selected flow rates.

Lower-priced units are not always the most economical. To determine the cost per inch of the UV lamp, the number of inches of ultraviolet lamps should be divided into the selling price of the unit. In the case of the four-lamp unit, we have 4 times 30 in., or a total of 120 in., which is divided into the unit's selling price. A typical unit of this size may cost around $1,300, or a cost of $10.75 per inch of UV lamp. In comparison, a larger unit costing approximately $1,800 typically consists of nine UV lamps, or 270 in. divided into the selling price of $1,800 to come up with a cost of only $6.65 per inch of UV lamp.

To complete a cost comparison of UV purifiers, it is necessary to determine the actual cost one is paying per cubic inch of UV purification chamber. This is accomplished by dividing the volume of the chamber into the cost. A competitive 50-gpm UV purifier has a chamber 6 in. in diameter and is about 30 in. long. A unit such as this would cost around $1.54/cu in. It is necessary to evaluate more information than just the manufacturer's rated capacity and the cost of the UV purifier.

TABLE 4.4 CHAMBER VOLUME AND LAMP NUMBER REQUIREMENTS

Flow Rate in gpm	Number of UV Lamps Required			Volume of Chamber or Chambers in U.S. Gallons
	30 in.	58 in.	60 in.	
25	5	3	3	6.25 gal
50	9	5	5	12.5 "
75	13	7	7	18.75 "
100	17	9	9	25 "
200	34	17	18	50 "
500	84	44	42	125 "
1,000	167	87	84	250 "

The following criteria should serve as a guideline:

- There should be a sufficient number of UV lamps to guarantee that the flow rate does not exceed two tenths of a gallon per minute per effective inch lamp (arc length).
- The volume of the purification chamber must be large enough to guarantee a minimum 15-second retention time.
- The UV unit should include a properly filtered UV sensor with appropriate alarm warning controls.
- The purifier must include a system to allow for cleaning the quartz jackets, sightport, and interior portion of the chamber without the need for disassembling the unit.

TABLE 4.5   TYPICAL TEST DATA FROM UV PILOT SYSTEM

MGD Flow	Fecal Coliform[a]		% Power To Unit	Suspended Solids
	Infl.	Effl.		
6.0	52,250	7	100	23
6.0	69,500	13	100	17
6.0	190,000	17	100	25
6.0	100,000	11	100	17
6.0	93,000	14	100	24
6.0	13,000	12	100	12
6.0	9,000	4	100	10
6.0	57,000	13	100	15
6.0	40,000	32	100	15
6.0	23,000	28	100	7
6.0	29,000	90	35	7
6.0	29,000	24	100	6
6.0	29,000	130	40	7
6.5	239,000	79	100	6
6.5	307,000	36	100	7
6.5	22,000	40	100	3
7.0	76,500	17	100	7
7.0	16,000	323	40	7
7.0	560,000	90	100	18
7.0	25,000	14	100	8
7.5	155,000	6	100	20
7.5	46,000	28	100	15
7.5	11,000	26	100	10
7.5	11,000	345	50	10
7.5	15,000	13	100	9

[a] Colonies per 100 ml (Millipore count). Note reduced power was required for analyses, not due to equipment failure.

- If a number of UV lamps are employed, the system must also include a lamp warning system which will sound an alarm in the event that one or more UV lamps go out even though the UV intensity may still be within a safe range.
- The true cost of the unit should be determined by calculating the cost per effective inch of UV lamp and the cost per cubic inch of UV exposure chamber.

Treatment of Secondary Wastewater Effluents

Despite the wide recognition of UV disinfection capability for many years, available equipment to utilize it was largely limited to the treatment of potable water. Various factors precluding its consideration as a practical treatment method for other fluids included unsatisfactory flow patterns, dead spots, or shadowing by internal mechanisms and hyperbolic deterioration of UV radiation effectiveness with increasing thickness of fluid films. However, UV disinfection has some rather significant economic advantages over chemical methods when additional cost items for the latter are considered. Removal of chlorine and its by-products and the use of ozone could more than triple the cost of existing secondary effluent treatment methods (from about 4.5 cents per 1,000 gallons to between 11 and 13 cents). In contrast UV treatment can cost as low as 1.5 cents per 1,000 gallons. Cost figures include both capital and operational items. Table 4.5 shows typical test data for a UV pilot system.

A typical commercial unit is illustrated in Figure 4.5. The unit is placed across the wastewater flow path to reduce liquid depth penetration to about

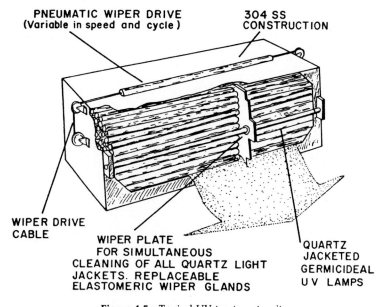

PNEUMATIC WIPER DRIVE (Variable in speed and cycle)

304 SS CONSTRUCTION

WIPER DRIVE CABLE

WIPER PLATE FOR SIMULTANEOUS CLEANING OF ALL QUARTZ LIGHT JACKETS. REPLACEABLE ELASTOMERIC WIPER GLANDS

QUARTZ JACKETED GERMICIDEAL U V LAMPS

Figure 4.5 Typical UV treatment unit.

6.3 mm maximum. Operating experience with these systems has been minimal.

Summary

Ultraviolet radiation is effective against waterborne pathogens. Its destructive action is fast and is typically more effective in clear water than halogens. No photoregeneration occurs under normal operation. The UV systems are compatible with gravity flow systems and the hydraulics of conventional plant designs. Also, UV does not react with chemicals in the water to form dangerous new compounds.

The disadvantages with UV are that there are no long-term residual germicidal effects. Color, turbidity, and iron can reduce the disinfectant potential of UV irradiation. Finally, these systems are generally more costly than other disinfectants.

5

Ethylene Oxide and Other Gaseous Sterilizers

Sterilization or disinfection by ethylene oxide (ETO) has been employed primarily in the sterilization of dairy products, disposable plastic items, and plastic bottles. Because of the complexity involved with this method and its particular limitations, sterilization by ethylene oxide is used only when other methods are not acceptable.

Ethylene oxide sterilizes by chemically reacting with (alkylating) biological compounds which are vital to normal metabolism and reproduction. Processes using ETO are more complicated than steam processes with cycles depending on temperature, exposure time, ETO concentration, and environmental humidity. A toxic substance, the use of ETO demands employee training in its proper use and handling, employee exposure monitoring, and employee medical surveillance. Sterilization by ETO has been used in 80 percent of the hospitals in the United States.

Ethylene oxide (ETO) is widely used in health-care facilities and industry to gas-sterilize materials that are sensitive to high temperature and moisture conditions. It is a method available to health-care facilities at this time for sterilizing medical devices that steam, dry heat, or other chemical sterilization techniques can damage. Ethylene oxide is not a substitute for steam sterilization, but it is an alternative method for temperature-sensitive items.

Ethylene oxide is a colorless gas at atmospheric temperature and pressure. Major physical properties are shown in Table 5.1. It is a highly active molecule readily reacting with chloride ions, water, acids, amines, and many

TABLE 5.1   MAJOR PHYSICAL PROPERTIES OF ETHYLENE OXIDE

Property	Value
Vapor Pressure at 68°F	1,094 mm Hg
Vapor Density (air = 1)	1.49
Boiling Point	51.3°F
Flash Point, Tag, Open Cup	<0°F
Flammable Limits in Air (% by volume)	
Lower Limit	3% (30,000 ppm)
Upper Limit	100%
Ignition Temperature	
In Air	804°F
In Absence of Air	1,060°F
Solubility in Water	Complete
Liquid Density (water = 1)	0.87
Color	Colorless
Odor Threshold	500–750 ppm

other hydrogen-containing materials. It is this same high degree of reactivity that makes ETO such an effective sterilant and conversely contributes to the chemical's human toxicity and flammability.

Ethylene oxide kills microorganisms in gas sterilization by a process known as alkylation. By definition, alkylation refers to the replacement of a hydrogen atom in a molecule of an organism with an alkali group ($-CH_2-CH_2OH$). In gas sterilization, ETO is the alkylating agent, which under the proper conditions of time, temperature, and humidity, can react with some of the cellular components of microorganisms. The result is cellular disruption and dysfunction. Typically the sterilizers used in health-care facilities maintain ETO concentrations of 600 mg/l to 1,200 mg/l.

Human effects resulting from gas exposures in excess of approximately 200 ppm include irritation of the eyes and respiratory passages. Headaches, nausea and vomiting, drowsiness, lack of coordination, weakness, and nervous disorders are some of the characteristic effects reported at higher concentrations.

Ethylene oxide is flammable in both its gaseous and liquid forms, although it is the vapors from the liquid which actually burn. Ethylene without an ignition source will only self-ignite in air if heated to a temperature of 800°F or above. This is the auto-ignition temperature which is the lowest temperature at which self-sustained combustion can occur in the absence of a spark or flame. Ethylene oxide without an air and ignition source will only ignite if heated to 1,060°F.

Gas sterilizers using 100 percent ETO are designed to remove the ignition source and air. These sterilizers operate their entire cycles under vacuum or negative pressure to draw out air from the inside of the chamber in the presence of ethylene oxide. It is important to note that nonflammable halogenated hydrocarbons can decompose to give off toxic and corrosive by-products at high temperatures. For this reason, they should not be used in the presence of flames, electric equipment, or other high-temperature sources.

ROLE IN INFECTION CONTROL

The increasing problem of hospital-related infection has focused attention on the vital importance of correct sterilization procedures.

Recommended sterilization assurance procedures include:

- Receiving, decontamination, cleaning, preparing, disinfecting, and sterilizing of reusable items.
- Assembly, wrapping, storage, distribution, and quality control of sterile equipment and medical supplies.
- Use of sterilization process monitors, including temperature and pressure recordings, and the use and frequency of appropriate chemical indicators and bacteriological spore tests for all sterilizers.
- Designation of the shelf life for each hospital-wrapped and sterilized medical item and, to the maximum degree possible, for each commercially prepared item, a specific expiration date that sets a limit on the number of days any item will be considered for use.
- The recall and disposal or reprocessing of outdated sterile supplies.
- Specific aeration requirements for each category of gas-sterilized items to eliminate the hazard of toxic residues.

Bacteria, their spores, and viruses are a major cause of infection and disease. Bacteria and viruses are capable of reproducing by millions at very rapid rates. Bacteria exist in what is termed an active or vegetative state, reproducing at regular intervals. They are also capable of survival under extremely adverse conditions by going into a protected dormant state, known as a spore. These spores can survive for extended periods, even under conditions of heat, cold, or desiccation which would kill vegetative bacteria and become active when again placed in a favorable environment. It is for this reason that sterilization must be an absolute process, with complete pathogen destruction.

Certain surgically critical bacteria can severely complicate health problems if not destroyed. For example, any surgical instrument that has only a few surviving bacteria or spores could become greatly contaminated, due to the rapid reproduction of the bacteria. Exposure to such pathogens could result in delayed patient recovery, infection, and even death.

Ethylene oxide sterilization is a complex process. In order for it to be an effective sterilizing agent, there must be the proper relationship between the gas concentration, moisture, time, and temperature. Alteration of any one of these four variables can affect the others and change the sterilization process. Though these conditions usually vary, based on the specifications of different sterilizer manufacturers, the basic principles are as follows.

In its pressurized container, ethylene oxide is a liquid and must be vaporized to effectively permeate and sterilize any load. Pure ethylene oxide gas is extremely explosive, flammable, and toxic. It is, therefore, often diluted. This can be mathematically calculated by using readings from the sterilizer pressure gauge. The range of effective concentrations is usually 450 mg/l to 1,500 mg/l. Higher concentrations usually result in a shorter sterilization time. This, however, is based on chamber temperature and other factors.

Moisture is measured in terms of relative humidity. This is the ratio of the amount of water actually present in the air to the greatest amount the air can hold at the same temperature. Ethylene oxide gas is generally considered a dry sterilization process. However, water vapor must be present, but not at the saturated (100 percent relative humidity) level of steam sterilization. The recommended level is at least 30 percent, since too much moisture can cause the formation of ethylene. This toxic compound can remain as a residue causing lethal action of ethylene oxide. Dry cells and spores are more readily oxidized by sterilization. With some sterilizers, the recommended procedure is to precondition the load by allowing the chamber to be filled with a high-humidity atmosphere for at least 30 minutes. This softens the surface of the spores and thus allows easier and faster penetration of ethylene oxide. With other sterilizers, moisture is added during the cycle. It is theorized that the water carries the ethylene oxide into the microorganism, where it can chemically react. Close attention should be paid to proper humidification, as it is probably the most common cause of nonsterility in gas sterilized items.

Exposure times can vary greatly. This is because each load is affected by the relative contamination, density, contents, and permeability to ethylene oxide gas. As a result, the sterilization must be adjusted accordingly.

Ethylene oxide vaporizes (from liquid to gas) at 51°F (10.5°C). It can be an effective sterilizing agent at temperatures as low as 70°F (21°C). Higher temperatures, however, allow shorter cycles by enhancing the gas diffusion rate. The usual operation temperature for ethylene oxide sterilizers varies from 70°F (21°C) to 140°F (60°C). Temperature also affects the pressure of the gas. Careful attention must be paid to the chamber temperature and gas pressure.

Ethylene oxide has provided the health-care industry with an important sterilization method for steam or dry-heat sterilization. Nevertheless, this sterilization is a slow process (much slower than equipment and a skilled operator). It also requires relatively expensive equipment and common problems include:

- Necessity to continually monitor temperature, gas concentration, and humidity on a manual sterilizer.

- Weakening of and or surface damage to certain acrylic and polystyrene items, the crazing of some lenses and dissolving of lens cement.
- Formation of toxic compound by-products.
- Retarding of gas penetration by protein matter and soil.
- The inability to sterilize solutions.

EQUIPMENT AND PROCEDURES

Various kinds of ethylene oxide sterilizers are in use today in semiautomatic and manual mode control systems. They range in temperature models from small to large, fully automatic built-in units.

The most commonly used steps of operation incorporate the following procedures:

Preparation. Materials should be surgically clean, free from soil or protein matter (for example, blood tissue or fluids), must be towel-dry and free from water droplets to minimize the formation of ethylene glycol. All cups, plugs, and valves should be removed to allow gas penetration. Needles and tubing should be open at both ends. Syringes should be disassembled.

Loading. Load similar to steam sterilizer. Gas should be allowed to circulate freely. Avoid dense packs.

Temperature. Preheat to operating temperature. This speeds up the lethal effect of gas.

First Vacuum. Drawing time varies with chamber size.

Moisture. A minimum relative humidity of 30 percent is needed for standard gas-sterilization cycles. Some sterilizers require the addition of water or wet sponges to supply the minimum moisture required. Automatic moisture injection into the chamber maintains the proper relative humidity. A conditioning period of 30 to 60 minutes should follow moisture injection. This time allows for humidification of dry bacterial spores and then permits rapid lethal action of gas.

Gas Injection. Ethylene oxide pressure and concentration vary with the sterilizer. Some sterilizers do not allow expansion and preheating prior to chamber entry. Preheating the gas eliminates expansion cooling effects and condensation of gas into liquid. Ethylene oxide/carbon dioxide mixtures tend to stratify in large storage cylinders due to their different atomic weights. Therefore, under certain conditions, the final 20 percent of the remaining cylinder contents may not contain sufficient ethylene oxide to be a sterilizing agent.

Exposure Time. It is imperative to operate according to the sterilizer manufacturer's instructions.

Second Vacuum. At end of the exposure time, ethylene oxide is removed from the chamber by drawing a second vacuum.

Atmospheric Restoration. At end of a cycle, filtered air is admitted to the chamber to restore atmospheric pressure.

Advantages and Disadvantages

The main advantage of ethylene oxide sterilization is its ability to sterilize those items which are adversely affected by steam or dry heat. Although ethylene oxide may sterilize most hospital items, a number of disadvantages limit its use. These include toxicity to personnel, higher costs, and longer exposure and aeration times. It is not recommended that ethylene oxide replace steam or dry-heat sterilization when the latter methods are applicable.

Ethylene oxide can be used to sterilize plastics, rubber, metal, leather, wood, wool, rayons, nylon, glass, and virtually every other material. Pure ethylene oxide does not cause damage to most materials. However, it can have adverse effects on some items when it is diluted with freon.

Aeration

Ethylene oxide gas provides an effective method of sterilization for heat and moisture-sensitive products. However, an excessive amount of ETO or its by-products may be harmful to patients and hospital personnel. Exposure to items such as prosthetic devices, surgical instruments, catheters, and so on which have not been properly aerated subsequent to sterilization could result in serious chemical burns as well as skin and mucous membrane irritation. Therefore, adequate aeration time must be allowed to ensure that any ethylene oxide and its by-products remaining in or on the sterilized devices have been reduced to a safe level.

SAFETY IN USING ETHYLENE OXIDE

Ethylene oxide is a strong, effective sterilant. If it can destroy microorganisms, it could be harmful to other living things as well. Users of ethylene oxide have long been cautioned to be careful when in areas that could be potential exposure areas.

STERILIZATION AND AERATION EQUIPMENT SHOULD BE

Installed correctly

Placed properly (in well-ventilated areas)

Located remotely

Maintained adequately

Studied for modification

Considered for replacement

EMPLOYEE'S WORK ENVIRONMENT SHOULD BE

Well ventilated
Away from sterilizers
Monitored periodically

EMPLOYEES SHOULD BE

Carefully trained
Constantly aware

SAFETY

For sterilizing in its pure form, pure ethylene oxide gas is extremely flammable.
Exercise caution.
Locate sterilizers in a restricted area to limit unauthorized access.
Locate sterilizers in a room that has between six and ten air changes per hour.
Always vent sterilizer exhaust to the outside.
After a complete cycle, open door with caution and clear a ten-foot radius around sterilizer.
Vent aeration exhaust to the outside.
Keep records of sterilizer malfunction and repairs.
Store tanks in special areas to meet building codes with tanks chained to an adequate support.

HEALTH PRACTICE

Avoid breathing ethylene oxide vapors.
Avoid direct skin contact.
Use protective gloves to take items out of sterilizer.
Minimize handling of sterilized items.
Report any accidents or prolonged exposure to supervisor.
Pull, don't push, loaded carts to the aerator.

Material Damage

Feron-diluted ethylene oxide can cause damage to certain plastic materials including tenite, styron, lucite, and plexiglas. It can cause crazing of some plastic lenses and dissolve some lens glues.

Ethylene Oxide Scrubbing System

Ethylene oxide scrubbing systems are designed for high-volume, high-efficiency, low-cost, and reliable operation. Scrubbers can be retrofitted on sterilizers or included in designing new sterilizers. Ethylene oxide from the sterilizer vacuum pump consists of a liquid, vapor, and low-moisture content reaction product of the scrubbing system, complete with the gas-dispersion system. This produces the most efficient and reliable ETO scrubbing system available. The liquid phase consists of an acid solution. The scrubbing solution permits repetitive use of the scrubber without having to regenerate the solution more than several times a year, depending on the amount of ETO discharged per cycle. Figure 5.1 is a schematic of a typical ETO scrubber system.

Gas Sterilization

Chemical vapors have a limited but useful role in the cold sterilization or disinfection of items of hospital equipment that are bulky or heat liable and in the industrial processing of some disposable articles. In a wider field, gas sterilization of component parts of space vehicles has been considered as a method of keeping the planets clean. The widely used gases at the present time are ethylene oxide and formaldehyde.

The biocidal activity of the various gaseous sterilants is related to their

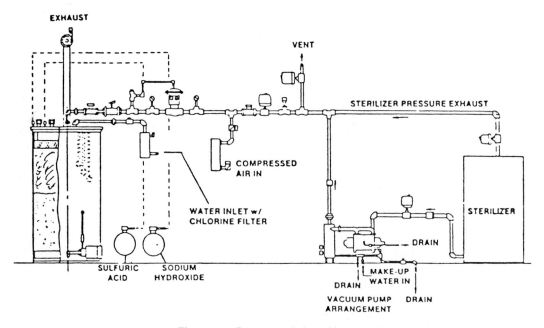

Figure 5.1 Recommended scrubber installation layout.

power of alkylating the SH, —OH, —COOH or —NH$_2$ groups of enzymes or other proteins and nucleic acids. The chemical basis of these reactions with ethylene oxide is as follows:

$$\text{protein—NH}_2 + \text{C}_2\text{H}_4\text{O} \Rightarrow \text{protein—NH—(C}_2\text{H}_4\text{OH)}$$

$$\text{protein—SH} + \text{C}_2\text{H}_4\text{O} \Rightarrow \text{protein—S—(C}_2\text{H}_4\text{OH)}$$

All types of organisms (including mycobacterium) are sensitive to these agents. Bacterial spores are only ten times more resistant than vegetative cells, compared to a ten-thousand-fold difference with respect to other chemicals.

ACCESSIBILITY OF ORGANISMS

Dry vegetative and sporing organisms are more resistant to ethylene oxide than those in a moist condition. This is a problem of some practical importance which is further complicated by the nature of the surface to be sterilized. For example, dried organisms on smooth, nonabsorbent surfaces such as glass, plastic, or metal are more difficult to kill than those on absorbent materials such as paper and textiles. This resistance may be due to a protective film of salt or protein around the organisms. On the other hand, susceptibility can be quickly restored by wetting, which redissolves protective films and releases organisms trapped in crystals. To prevent organisms from becoming trapped in crystalline materials, it is necessary to sterilize the mother liquor by filtration prior to crystallizing the product aseptically.

There is an optional level of relative humidity in gas sterilization which is about 33 percent of ethylene oxide and 80 percent to 90 percent for formaldehyde and B-propiolactone. In the case of ethylene oxide, an increase in relative humidity slows the rate of kill, while a decrease may cause actual failure to sterilize. Occasional failures, referred to as skips, occur when the relative humidity is dropped to 22 percent and are due to the resistance of dried organisms.

The effective concentration is the partial pressure, that is, the actual amount of the gas in the sterilizer and not its concentration relative to the inert gases used as diluents. The concentration, which determines the rate of sterilization, may be reduced by absorption: for example, blankets absorb 1 percent of their weight of formaldehyde, while rubber absorbs 15 mg. of ethylene oxide per gram, and plastics also have a high absorptive capacity.

APPLICATIONS

Gas sterilization, in particular with ethylene oxide, is now used on an extensive scale but in many industrial processes it is likely to be replaced in the future by gamma radiation.

The advantages of gas sterilization are:

- Articles are not damaged by the process.
- Materials may be treated at low temperature in an air-dry state.
- Penetration of gas into porous materials is usually good, with the exception of formaldehyde.

Disadvantages are:

- Plastic wrapping must be left open and sealed after treatment.
- Control of the relative humidity and hydration of the organisms is critical.
- Sterilization time is usually long.
- Cost of processing is higher than that of heat sterilization.

OTHER GASEOUS STERILANTS

Propylene Oxide

Propylene oxide is less effective than ethylene oxide but has the advantage that it is liquid at room temperature and is less explosive. Thus, it can be used safely in a pure state for small office or laboratory sterilizers. A small

TABLE 5.2   PROPERTIES OF GASES USED IN STERILIZATION

Property	Ethylene Oxide	Propylene Oxide	Formaldehyde	β-Propiolactone
Formula	$H_2C\!-\!CH_2$ with O bridging (epoxide)	$H_2C\!-\!CH\!-\!CH_3$ with O bridging (epoxide)	HCHO	$CH_3\!-\!CH_2$ over $O\!-\!C\!=\!O$
Boiling point	10.7°C	34°C	90°C (from solution)	163°C
Inflammability	Explosive	Moderate	None	None
Penetrating power	High	High	Low	Low
Toxicity	Irritant Vesicant	Irritant Vesicant	Irritant Pungent	Irritant Vesicant ?Carcinogenic
Biocidal concentration, mg. per litre at 20°C	400–1,000	800–2,000	2–5[a]	2–5
Optimal relative humidity	33%	33%	>75%	>75%

[a] This is the highest concentration attainable at low temperature; not optimal concentration.

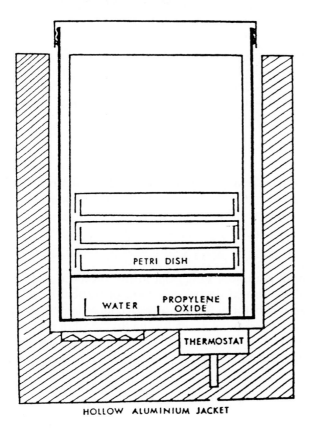

PETRI DISH

WATER | PROPYLENE OXIDE

THERMOSTAT

HOLLOW ALUMINIUM JACKET

Figure 5.2 Propylene oxide disinfector designed for dental use.

sterilizer (Figure 5.2) is shown to meet some of the problems of office sterilization in dental and medical practices. With this apparatus it is possible to sterilize articles that cannot conveniently be treated by moist heat, such as hand towels, swabs, dressings, and small instruments excluding syringes and needles.

A rise of 10°C may increase the rate of sterilization threefold, as in the case of ethylene oxide, which is used over a range of temperatures from 20°C to 56°C. Formaldehyde also has a high-temperature coefficient and this property is used to increase its effectiveness in the disinfection of blankets. Some of these properties of the gaseous sterilants are given in Table 5.2.

Beta-Propiolactone (BPL)

This liquid has a high virucidal and bactericidal activity. It has a high boiling point but can be vaporized by means of an atomizer and is reported to be superior to formaldehyde for the disinfection of rooms. It has been demonstrated to be carcinogenic when applied to the skin of mice.

Glycols

Ethylene and propylene glycols in vapor form have been used with doubtful success as aerial bactericides in hospital wards and in school rooms. A bactericidal concentration, 0.2 to 0.5 mg per liter, can be produced by heating 50 to 60 ml per 1,000 cubic feet of enclosed space. The optimal relative humidity is 60 percent. These are not widely used.

Chloropicrin

This gas Cl_2CNO_2, having bactericidal properties, has found some application as a soil-fumigant, where even if complete sterilization of the soil is not achieved, many phytopathological fungi are destroyed.

Methyl Alcohol

The vapor of this alcohol is rapidly fungicidal in the presence of high humidity. It can be used for fungal decontamination of incubators and cold rooms. A standard-size incubator at 37°C can be decontaminated overnight by placing it in a pad of cotton wool soaked in methyl alcohol and another pad soaked in 20 ml of water. As the vapor is inflammable, care must be taken when opening the incubator door that there are no naked flames in the area. It is toxic.

Methyl Bromide

This compound has a low antimicrobial activity requiring a concentration of 3.5 g per liter and a relative humidity between 40 percent to 60 percent. The advantages claimed for methyl bromide are its high penetrating power and low toxicity. It has been considered in food processing in place of ethylene oxide, and it is used as an insecticide and weed killer in agriculture.

Ethylene Imine

$$CH_2\text{---}CH_2$$
$$N$$
$$\|$$
$$H$$

This gas is more active than ethylene oxide under certain conditions (such as high relative humidity) and has more than 100 times the activity of ethylene oxide. Like formaldehyde, it is markedly dependent upon high relative humidity. Moreover, it is both flammable and corrosive, particularly to many metals. Its high order of activity may perhaps permit it to be used at concentrations low enough so that these factors are not particularly objectionable, but to date it has been put to little or no practical application as a vapor-phase sterilant, although it has been used in liquid solution in vaccine production, including that for foot and mouth diseases.

Peracetic Acid

$$CH_3-\overset{\overset{\displaystyle O}{\|}}{C}OOH$$

Considerable application, both as an aqueous solution and as an aerosol or vapor in germ-free environments or research, has been found for peracetic acid, where its corrosiveness as both an acid and a strong oxidizing agent is circumvented by the use of corrosion-resistant materials for animal isolators and other equipment. Its advantages lie in its quick action, even against usually resistant spores, and the fact that its decomposition product is nontoxic (volatile acetic acid). Caution must be exercised in keeping the compound, or solutions of it, too long in tightly closed containers where the pressure build-up by released oxygen may cause them to rupture or explode. While highly active it requires a high relative humidity, 80 percent being optimal, and its penetration properties are poor. Spores on porous surfaces such as paper are more readily attacked than those on impervious glass surfaces.

6

Electrolysis ━━━━━━

Electrolysis is a chemical process by which chemical reactions are produced electrically either in solutions or molten salts. The most important industrial applications of electrolysis, both existing or potential, fall into three broad categories: metal recovery and electroextraction; electrochemical organic synthesis; and electroconcentration of solids, cleaning and disinfection of liquids.

Electrolysis metal recovery from waste effluents is carried out on a large and commercial scale in the copper industry on pickle liquors. New electrolysis processes have been developed and are also in use for extraction of metals from sulfide ores; these processes avoid the SO_2 evolution which takes place in the traditional melting process. A vast number of elements and inorganic compounds are produced by electrolytic methods including chlorine, sodium, magnesium, lithium, nickel, zinc, caustic potash, soda ash, and sodium chlorate.

Commercial organic synthesis processes are another application of electrolysis. There is a potential for expansion of this application because electrolysis offers greater product selectivity, lower temperatures, reduced pollution, and alternate starting materials. Examples of industrial processes which have been in use include the manufacture of sorbitol adiponitrile, tetramethyl lead, and fluorocarbons. In electroconcentration and disinfection, electroflotation processes have been applied to paint-bearing wastewaters and to steel-rolling mill and paper mill effluents. Table 6.1 shows examples of electrolyte processes.

TABLE 6.1 EXAMPLES OF ELECTROLYTIC PROCESSES

Area	Electrolysis	Alternate Processes
Cyanide Waste Treatment Cynox Process	Electrolysis	Alkaline Chlorination
Electrowinning Copper	Electrolysis	Liquid Ion Exchange
Zinc		Smelting Furnace
Propylene Oxide Manufacturing		
Cyclohexadiene Production		
Hydroquinone Synthesis	Electrochemical	Chemical
Acid Mine Drainage	Electrolysis	Lime Treatment
Cheese Whey	Electrolysis	Biological Systems
Oil Refining Edible Fat	Electrolysis	Skimmer and Air Flotation
Brass Wire Mill Copper Recovery	Electrolysis	Pickling Waste
Domestic Waste Sterilization	Electrolysis of NaCl to Produce Hypochlorite	Chlorination
Metal Recovery	Melectrolytic Cells	Precipitation and
Material Separation	Electroflotation	Settling Tanks
Inorganic Production		
Chlorine		
Caustic Potash		
Soda Ash		
Sodium Chlorate		
Na		
Al		
Mg		
Ti		
Li		
Cu		
Ni		
Zn		
Hg	Electrolysis from Sulfide	
Mo	Electrolysis from Sulfide	
Ag	Electrolysis from Sulfide	

Applications of electrolysis derived from research and development programs which have been set up by the industry include the synthesis of compounds such as propylene oxide, melamine, phenols, quinones, and organic acids. Another broad area is the electroconcentration of solids and cleaning of liquids, especially for treating acid mine drainage waters, cheese whey, chromated waters, cyanide wastes, domestic sewage, and organic wastes. In general, the electrolytic processes lead to less pollution than many conventional processes, and yet also compare well economically with conventional processes.

PRINCIPLES AND TECHNOLOGY

Electrolysis is a method by which chemical reactions are produced electrically in solutions of electrolytes or in molten salts. The basic reactions, which occur at the electrodes immersed in these solutions, are reduction and oxidation. Reduction involves the loss of electrons at the cathode, and oxidation involves the gain of electrons at the anode. An electrolytic cell system is shown in Figure 6.1. In this system, the electrolytes are cooled by external heat exchangers.

Applications of electrolysis utilizing electricity to break down molecules cover a broad spectrum of industrial processes. They differ by the size of the equipment employed and the unit power rating, ranging from domestic sewage sterilization to production of commodities such as chlorine (10 million tons per year). Included also are such uses as metal recovery from industrial wastewaters and synthesis of organic compounds.

Broad divisions are used to categorize the applications of electrolysis, whether they are existing or potential (in the research stage): metal recovery

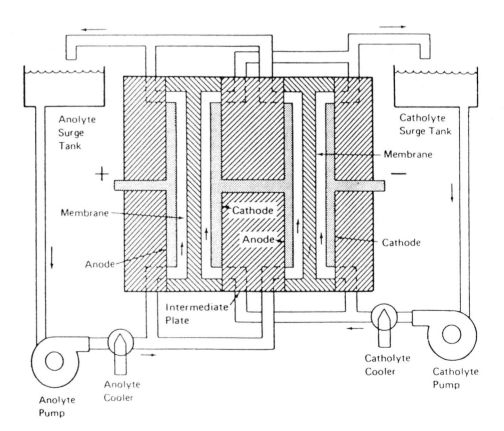

Figure 6.1   Schematic of an electrolytic cell system.

and electroextraction; electrochemical organic synthesis and electroconcentration of solids; cleaning and disinfection of liquids. These areas cover the major uses of electrolysis in industry and the uses which are most important because they have the largest potential for future development.

The quantity of electrolysis products, their rate of production, and quite often their nature depend on electrolysis conditions. According to Faraday's Law, the quantity of substance being consumed or produced by a single electrode reaction is proportional to the quantity of electricity consumed in the electrolysis. This quantity of electricity is equal to the product of the current multiplied by the duration of electrolysis for a constant current or to the integral of the current over the duration of electrolysis for a variable current. Table 6.2 shows the variables of concern in any chemical reaction as well as those variables peculiar to electrochemistry.

The nature and relative abundance of electrolysis products at each electrode generally depend on the electrode potential. Control of the electrode potential is accomplished by two means: selection of an electrode possessing the appropriate overvoltage potential, and use of a high overvoltage electrode, controlling the applied voltage to maintain the desired potential. Overvoltage is the potential difference between that required for gas evolution at the electrode and that potential corresponding to equilibrium for a reversible system. Each different type of metal has unique overvoltage. Overvoltage is influenced by such factors as current density, temperature, pressure, electrode history, and electrolyte pH.

In order to govern the rate of an electrochemical reaction, two factors should be known in addition to electrode potential: specific rate constants and mechanisms for the overall change-transfer process at the electrode, and the mass-transfer rates of species at the electrode surface. With the exception

TABLE 6.2   VARIABLES IN ELECTROCHEMICAL REACTIONS

Usual Variables	Electrochemical Variables
We are normally concerned with these variables in chemical reactions.	An electrochemical reaction requires knowledge of both the usual and the electrochemical variables.
Concentrations of:	Electrode potential
(1) Solvent/supporting electrolyte	Electrode material
(2) Substrate	Current density
(3) Other components:	Electric field
Temperature	Adsorption
Pressure	Cell design
pH	(1) Compartmented?
	(2) Membrane or porous separator?
	(3) Static or flow cell?
	Solution conductivity

of electrode potential, these factors bear a direct analogy to conventional design parameters for heterogeneous reactions.

APPLICATIONS

Currently, the most widespread commercial applications of electrolysis are metal recovery, particularly copper; electrooxidation of sulfide ores; the large-scale production of inorganics, particularly chlorine and sodium; the synthesis of organic compounds; and electroflotation to transform slurries into clean liquids.

The incentives for metal recovery are the money saved by recovering the metal itself and the simplification of wastewater disposal. The most common application of metal recovery by electrochemical methods is found in the copper refining industry. Freshly rolled, drawn, or fabricated copper is pickled in a sulfuric acid tank, dissolving the scale, copper oxide, which converts the acid into sulfate of copper. In time, the pickle liquor loses strength. Historically, this was either discharged to waste or chemically worked up into copper sulfate. Now, however, all modern plants are equipped with electrolysis units which electrolyze the copper sulfate solution and convert it to metallic copper and sulfuric acid, with oxygen evolved at the anode.

The copper recovery system is a closed loop as shown in Figure 6.2. The pickle liquor goes from pickle tank to recovery unit and back again. While copper sulfate must be converted back to acid and copper at each pass through the recovery cell, there is no need for the level of $CuSO_4$ to be reduced to low levels. This means that the copper recovery takes place in a solution of relatively high copper concentration, with the effect of high efficiency of removal. Copper recovery plants are typically under the conditions shown in Table 6.3.

TABLE 6.3   OPERATING CONDITIONS FOR
COPPER RECOVERY CELL

Anodes	PbO_2 (1–2% Ag)
Cathodes	Stainless steel or copper
Cell temperature	48–65°C
Current density	4–16 mA/cm^2
Cell voltage	2.5–3.5 V
Cu concn. (in)	7%
(out)	2%
Efficiency (C.E.)	70–90 m
Electrode gap	1–4 cm
Energy expenditure	900 AH–4.5 kWh/kg Cu

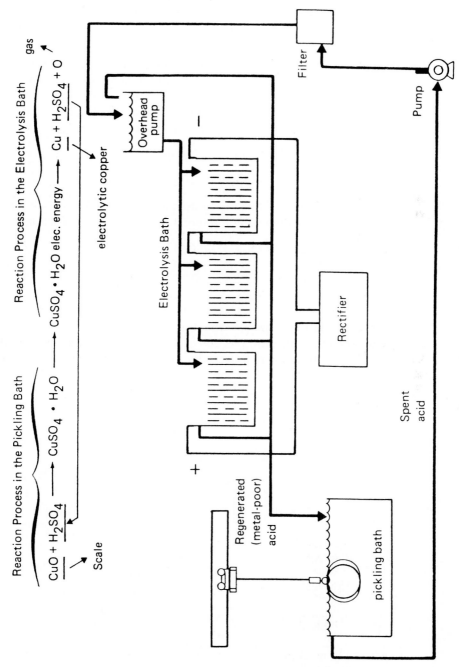

Reaction Process in the Pickling Bath

$$CuO + H_2SO_4 \longrightarrow CuSO_4 \cdot H_2O$$

Reaction Process in the Electrolysis Bath

$$CuSO_4 \cdot H_2O \xrightarrow{\text{elec. energy}} Cu + H_2SO_4 + O$$

gas

electrolytic copper

Scale

Electrolysis Bath

Overhead pump

Rectifier

Filter

Pump

Spent acid

Regenerated (metal-poor) acid

pickling bath

Figure 6.2 Plant flow sheet for copper removal process.

ELECTROOXIDATION PROCEDURES FOR THE RECOVERY OF SULFIDE ORES

Recovering sulfide ores by electrooxidation involves crushing and grinding of the ore, forming a pulp with brine solution containing sodium chloride, and electrolyzing the pulp. Oxidation of sulfide minerals occurs as a result of the controlled production of hypochlorite during electrolysis of the brine solution containing the finely ground ore. The anode reactions involved in forming the oxidizing species are shown in the equations:

$$2\,Cl^- - 2e \rightarrow Cl_2$$

$$Cl_2 + H_2O \rightarrow OCl_1^- + 2H^+ + Cl^-$$

The hypochlorite ion (OCl^-) formed as a result of the anode reactions oxidizes or chlorinates the sulfide mineral according to the reactions:

$$MS + 4\,OCl \rightarrow MSO_4 + 4\,Cl^-$$

or

$$MS + 4\,OCL \rightarrow MCl + SO_4^{2-} + 3\,Cl^-$$

where M refers to the metal contained in the mineral.

With mercury, for example, the sulfate reacts with the chlorine in the electrolyte to form the soluble tetrachloro mercury complex. For silver, the

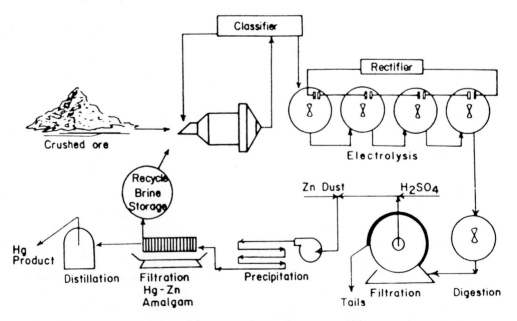

Figure 6.3 Flow diagram for electrooxidation of mercury sulfides or similar ones.

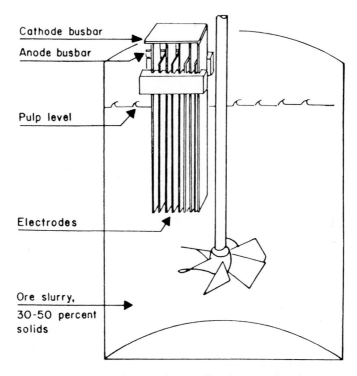

Figure 6.4 Electrooxidation cell and agitator vessel.

soluble tetrachloro complex $AgCL^{3-}$ is formed. For molybdenum, the soluble molybdate ion is formed. For copper, the chlorocomplex $CuCl^-$ is formed, and for antimony, $Sb_2O5—$ is formed.

Electrodes used may be iron for the cathode and PbO^{2-} coated titanium anode, or an all-graphite system. Salt percentage in the solution lies in the 3 percent to 10 percent range. The pH of the electrolyte is allowed to follow the natural pH obtained on contact of solution with the ore, which usually falls in the pH 6 to 7 range (effective dissolution of the molybdate ion requires pH values above 7.5).

Figure 6.3 shows the flow diagram and Figure 6.4 illustrates the cell and agitation vessel for this process. Agitation of the pulp was necessary during electrolysis so that the oxidizing species produced at the anode was effectively mixed with the mineral particles for the chemical reaction. The electrodes were immersed in the pulp and positioned so that the agitator continuously forced the pulp between the electrodes. The hypochlorite ion reacted as rapidly as it was formed in the agitated cell, thus maintaining hypochlorite at a low concentration so that undesirable oxidation reactions were minimized. Power consumption for electrolysis is given in Table 6.4. The optimum operating temperature of the solution is in the range of 25°C to 40°C to ensure both a good reaction rate (increasing with temperature), and a good stability

TABLE 6.4 POWER CONSUMPTION FOR ELECTROOXIDATION OF SULFIDE ORES (PER TON OF ORE)

Metal Extracted	Energy (kWh/ton)	Remarks
Mercury	17–60	1–4 lb Hg/ton ore
Molybdenum	160	From molybdenite 20 kWh/lb molybdenum recovered
Silver	40–70	20% salt concentration required

of the hypochlorite solution (decreasing with temperature). Retention time in the electrolytic cells was about 4 to 6 hours.

Recovery of metal values from chloride solutions is accomplished by conventional means. In the case of mercury, precipitation is by contact with a mesh or metallic mercury and in pilot plants was 94 percent.

Zinc minerals, nickel, cobalt-sulfide ores, and certain copper and antimony ores containing sulfide minerals might be amenable to the electrooxidation process. Potential also exists for recovering metal values from ores containing organic materials that inhibit dissolutions of metals by more conventional leaching methods and may otherwise require a roasting treatment.

In addition, electrooxidation, as an alternate to smelting of sulfide ores, avoids emission of SO_2 gases to the environment. Product sulfur, in the form of sulfate, reacts with gangue constituents and is discharged in the tailings.

ORGANIC SYNTHESIS PROCESSES

With the impetus coming from fuel cell research, the economic and technical attractiveness of possible electroorganic synthesis is being realized. The advantages of electroorganic synthesis over conventional processes are:

1. Many reactions may take place at controlled rates and at comparatively great product selectivity.
2. Process conditions are usually simpler than reaction by conventional means (for example, high temperature may be avoided).
3. The potential exists for reduced air and water pollution.
4. Alternate starting materials and new synthetic routes to synthesizing molecules may be exploitable, thus giving the chemical industry a greater diversity.

Conversion of Glucose to Sorbitol

One of the first electroorganic reductions to achieve commercial significance was the conversion of glucose to sorbitol, developed by Atlas Powder Company in the mid-1930s. Prior to that time, sorbitol was obtained from fruits and other plants, where it was present in very small concentrations and sold for several hundred dollars per pound. The Atlas electrolytic process for sorbitol involved the following cathode reaction:

$$
\begin{array}{l}
\text{CH}_2\text{OH} \\
\quad| \\
\text{HO—C—H} \\
\text{HO—C—H} + 2\text{H}^+ + 2e^- \\
\text{H—C—OH} \\
\text{HO—C—H} \\
\quad| \\
\text{CHO}
\end{array}
\rightarrow
\begin{array}{l}
\text{CH}_2\text{OH} \\
\quad| \\
\text{HO—C—H} \\
\text{HO—C—H} \\
\text{H—C—OH} \\
\text{HO—C—H} \\
\quad| \\
\text{CH}_2\text{OH}
\end{array}
$$

d-Glucose (Dextrose)   d-Sorbitol

The commercial cell employed a pure lead anode and a lead-mercury amalgam cathode. The Atlas sorbitol plant, built in 1937, had a capacity of 3 million lb/year of sorbitol, mannitol, and related compounds. This early commercial success for electrolytic processing of organic chemicals was short-lived. Within a few years it had been replaced by a high-pressure catalytic hydrogenation process.

Adiponitrile, Tetraethyl, and Tetramethyl Lead

In the organic sector the most significant commercial applications are for adiponitrile and tetraethyl and tetramethyl lead production. (See Figure 6.5.) For over a decade the Monsanto Company has been producing several million pounds per year of adiponitrile from acrylonitrile. Capacity of this process increased somewhat in the late 1960s. Asahi in Japan also commercialized a similar process.

Adiponitrile (Monsanto)

$$
2\ \text{CH}_2 = \text{CHCN} \xrightarrow{\text{2F}} \text{N} \equiv \text{C} - (\text{CH}_2)_4 - \text{C} \equiv \text{N}
$$
$$
\text{(ACN)} \qquad\qquad\qquad \text{(ADN)}
$$

Tetraethyl Lead (Nalco)

$$
\text{Pb} + 4\ (\text{C}_2\text{H}_5)\ \text{MgCl} \xrightarrow{\text{4F}} \text{Pb}\ (\text{C}_2\text{H}_5)_4 + 4\text{Mg Cl}^+
$$
$$
\text{(TEL)}
$$

Figure 6.5 Commercial electroorganic syntheses.

ELECTROFLOTATION

Electroflotation is employed to separate dilute suspensions into slurries and clear liquid. Electrolytically generated bubbles effect a separation by rising up through a suspension or colloidal liquid. The electrodes used in this process may be platinized titanium, carbon steel, stainless steel, or lead dioxide. Alternative methods utilize settling tanks or flocculating agents, but settling tanks are costly and occupy valuable land space, and flocculating agents require dosing and metering equipment.

Potential uses of electrolysis for treating wastewater, defined as those that have not received widespread usage, include:

- Electrochemical removal of heavy metals from acid mine drainage; reducing the chemical oxygen demand (COD) of cheese whey waste generated from dairy processing plants.
- Recovery of fatty materials from edible fat and oil refining plants.
- Regeneration of chromated aluminum deoxidizers (metal-finishing wastes).
- Electric treatment of domestic water.
- Cyanide waste treatment.
- Electrochemical destruction of organics.
- Electroflotation.

ACID MINE DRAINAGE

An acid drainage problem occurs when, in the process of mining coal, the surrounding rock strata are exposed to air and water. These rock strata contain metal sulfides, the most common of which is iron pyrite (FeS_2). The sulfides of aluminum (Al_2S_3) and manganese (MnS) are also found to a lesser extent. When in contact with air and moisture, these sulfides are oxidized to sulfates via a reaction that also generates acid. The basic reaction involving iron pyrite is:

$$2\ FeS_2 + 2\ H_2O + 7\ O_2 \quad 2\ FeSO_4 + 2\ H_2SO_4$$

The drainage from coal mines will vary in the degree of contamination from area to area. A typical acid mine drainage will be composed of:

Fe^{+2}: 100–500 ppm SO_4: 1,000–4,000 ppm

Fe^{+3}: 90–300 ppm Mn^{+2}: 15 ppm pH = 2.9–4.0

Various physical-chemical processes for the removal of contaminants from acid mine drainage have been considered. Purification by crystallization, distillation, ion exchange, and reverse osmosis produce high-quality water.

However, it must be recognized that each of these techniques is not a complete treatment. In addition to pure water, there is also produced a concentrate of the contaminating salts and acid which subsequently must be treated by chemical means before disposal.

Electrolysis has been considered as a technique for oxidizing ferrous iron in acid mine drainage and for removing total iron by plating and precipitation. However, standard electrolytic methods have not proven practical when applied to the removal of trace contaminants from water, for example, less than 1,000 ppm. The long retention time required for the ions to migrate to the electrode surface, together with the high-power consumption necessary for treating the increasingly purer water, have seriously restricted the use of electrochemical technology. Conventional electrolysis has been used generally only in waste treatment cases where high concentrations are reduced to lower concentrations prior to the chemical treatment.

Electrolytic cells which contain a semiconductive bed of particles between the electrodes, while not reacting chemically with the contaminants in solution, behave as small intermediate electrodes in the electric field between the major electrodes; that is, ends of the particles themselves become bipolar. The result is a cell with a multiplicity of anodic and cathodic sites. The cell can also be modified to provide a bed with predominantly anodic or cathodic character. When the solution to be treated is passed through this medium, the distance the contaminant ions have to migrate to any one of the numerous electrode sites is drastically reduced. This feature overcomes the excessive residence time required by conventional cells in the treatment and, because of their proximity in the cell, abnormally high voltages are not needed to oxidize or reduce trace quantities of dissolved impurities.

The cells are designed in two different configurations. A typical rectangular cell consists of a stainless steel cathode and graphite anode. Particulate electrode bed media consist of carbon. The cells run at 8–25 V and 3 A. The diaphragm permits operation of the cell with the bed particles isolated from either anode.

Cost savings from electrolysis include:

- Neutralization with cheaper limestone rather than lime.
- A reduction in sludge settling time due to the better properties of limestone sludges.
- Reduction of sludge disposal volume.

CHEESE WHEY

Whey from dairy processing plants constitutes a significant source of water pollution in many areas of the United States. The wastes generated by these plants may be characterized either as concentrated wastes with high solids concentration and biochemical oxygen demand (BOD), or as dilute wastes which are obtained in rinsing cures and cleaning equipment. The problems in handling these two general types of waste are quite different.

Whey remains as the liquid fraction when milk is curdled and the curds are separated by screening. These curds contain most of the casein and fat while the whey contains the lactose, salts, albumin, and globulin as well as acid substances, such as lactic acid, which assist the curdling process. Characteristics of the cheese depend on the conditions of curd formation.

Whenever economic circumstances permit, it is desirable to recover the value in these wastes. For example, products for human or animal nutrition are often produced by concentrating and drying. To a considerable extent, the recovery of whey solids as dried whole whey and as specialty food products is being practiced. One such specialty is partly desalted whey solids for infant feeding.

REGENERATION OF CHROMATED ALUMINUM DEOXIDIZERS

Chromium-containing chemical compounds have long been recognized as a major contributor to water pollution problems because of the high toxicity of chromium's ionic forms and the prevalence of chromium in a wide variety of industrial processes. Approximately 10 percent of the total U.S. chromium consumption is in the chemical industries (in excess of 50,000 tons annually). Nearly one-third of this is used for paint pigments and hence can be considered as nonpolluting, as are the metallurgical and refractory uses.

Metal surface treatment and corrosion control measures also use a large quantity of chromium, estimated at 30,000 tons annually. In this metal-finishing industry, chromated processing solutions are used extensively to treat aluminum surfaces during various operations such as anodizing, conversion coatings, prepaint preparation, welding, and adhesive bonding. A specific process commonly referred to as deoxidizing of aluminum (a part of a cleaning cycle) is of special interest. Chromated aluminum deoxidizing solutions have a relatively high concentration of chromium in the hexavalent state, and this chromium is used up in three ways: (1) a minute amount remains on the surface of the aluminum as a complex chemical conversion coating; (2) a somewhat larger amount is lost by drag-out into rinse waters; and (3) high concentrations are lost when the processing solution is discarded for various nonfunctional reasons. It is predicted that for many technical and economic reasons, chromated aluminum deoxidizers will continue to be used.

The regeneration concept involves oxidation of trivalent (depleted) chromium (Cr^{+3}) to hexavalent ($Cr_2O_7^{-2}$) (active) chromium at an electrode. This step requires that an electrical circuit be maintained within the solution, yet movement of dissolved metals must be restricted to specific areas. Infinite life span of the solution also implies separation and removal of aluminum and other metals that are dissolved in the deoxidizing process.

ELECTROLYTIC TREATMENT OF DOMESTIC WASTES

Sterilization of domestic sewage by electrolysis has been set forth. There have been two different methods of approach for this problem. In one concept, developed at the beginning of this century, wastes were electrolyzed as they arrived in the plant, with lime addition to control pH. In the second method, the electrolysis of NaCl generated hypochlorite which then sterilized the sewage. In earlier versions, this was applied by mixing sewage and sodium chloride, and then electrolyzing the mixture. It was later appreciated that an electrolysis of the NaCl alone, with subsequent discharge of the chlorine-rich liquid into a mixing tank where it reacted with the sewage, avoided problems associated with fouling of electrodes. A flow sheet for electrolytic sewage treatment is illustrated in Figure 6.6.

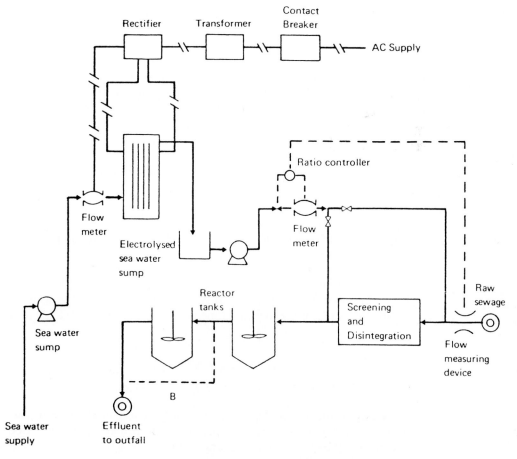

Figure 6.6 Flow sheet for electrolytic sewage treatment.

The first of these two methods is of historical interest during the first quarter of this century. This method treated raw sewage with lime and electrolyzed it using iron electrodes. The solids were allowed to settle and the effluent discharged. The effluent was described as being clear, odorless, and nonputrescent. Several demonstration plants were built and became the center of considerable controversy with doubt expressed about the effectiveness of the method; some critics questioned whether the system worked at all. It was claimed that the lime treatment and settling produced the same result whether or not the sewage was electrolyzed. After the closing of plants at Santa Monica, California, and Oklahoma City, Oklahoma, in the 1930s, no further installations were constructed. No large-scale electrolysis installations have been installed at municipal plants because of their reputation for high operating and capital costs.

CYANIDE WASTE TREATMENT

Most of the cyanide wastes which cause stream pollution problems in waste treatment plant operation are in the discharge wastes of industrial processes which include metal plating, case hardening of steel, neutralizing of acid "pickle" scum, refining of gold and silver ores, and scrubbing of stack gases from blast and producer gas furnaces. The greatest source of cyanide-bearing waste comes from the rinse waters, spillages, and drippings from the electroplating solutions of cadmium copper, silver, gold, and zinc. Previously, the standard waste treatment technique for this application was the alkaline chlorination process. In this process, the following reactions occur when chlorine is injected into the wastewater.

$$Cl_2 + CN^- \rightarrow CNCl + Cl \text{ (for free cyanide)}$$

This reaction is practically instantaneous and its rate is independent of pH:

$$Cl_2 + H_2O \rightarrow HOCl + HCl$$

Cyanogen chloride hydrolysis is a fairly strong function of pH. At a pH of 10 to 11, the time for 100 percent conversion to occur is on the order of 5 to 7 minutes. At a pH of 8.5 to 9, the time is increased to from 10 to 30 minutes.

$$2\,CNO^- + 4\,H_2O \xrightarrow{Cl_2} (NHH_4)_2CO_2 + CO_3$$

$$3\,Cl_2 + (NH_4)_2CO_3 + CO_3^{2-} + 6\,OH \rightarrow$$

$$2\,HCO_3^- + N_2 \uparrow + 6\,Cl^- + 6\,H_2O$$

This reaction finally completes the destruction of cyanide to the final products of nitrogen gas and the bicarbonate ion. If the pH is lowered, the bicarbonate will decompose and liberate CO_2. During the addition of chlorine to an aqueous cyanide system, the pH drops markedly. To overcome this problem, caustic must be added and takes part in the CNCl reaction as well as in the final reaction where nitrogen is evolved. The pH effect on chloride hydrolysis is:

$$Cl_2 + H_2O \quad HOCl + HCl$$

Problems encountered in the operation and maintenance of an alkaline chlorination system such as chlorine gas leaks, sensor problem failure, and carbonate scaling, plus the costs of chemical raw materials required by the system, make it desirable to develop alternative means for the destruction of cyanides in wastewater. In addition new and more restrictive regulations regarding cyanide discharge have been promulgated.

Other approaches are possible which emphasize recovery of either metal or cyanide. Cyanide may be reduced by electrolytic oxidation of concentrated baths to as low as 5 ppm, but reduction below 1,000 ppm is difficult. Usually several hours to several days are required to complete the reaction. A temperature of 200°F is needed and about 3.5 kWh are required per pound of cyanide destroyed. In addition, residual cyanide left in the bath must be destroyed by conventional methods.

Evaporation is another technique that has been used to recover chemicals from plating baths. For example, vacuum evaporators have been proposed for cadmium-cyanide rinse waters with the concentrated cyanide being returned to the plating bath and the distillate being reused for rinse. Evaporators are expensive and difficult to maintain. Entrainment of cyanide in distillate is also a problem.

Reverse osmosis has been successfully employed on nickel rinse waters, and is also applicable to rinse waters from copper or zinc cyanide baths. Rinse waters containing 800 ppm sodium cyanide can be concentrated more than threefold by reverse osmosis with removal levels of over 98 percent copper and zinc and over 92 percent sodium cyanide. Effluent water from the reverse osmosis unit could be reused to rinse and the concentrate would be returned to the plating bath for makeup capital.

Anodic oxidation offers potential solutions to the destruction of a variety of compounds. By electrolytic techniques analogous to those utilized for cyanide destruction, acetate wastes, phenols, cresols, cyanates, and thiocyanates may be treated. The main problem behind these treatments lies with process inefficiencies at low concentrations due to poor mass-transfer conditions.

ELECTROFLOTATION

Effluents which might typically be treated by electroflotation include:

Oil industry wastes
Engineering industry wastes (cutting oils, and so on)

Slaughterhouse wastes
Food industry waste (vegetable-animal oils)
Dairy wastes (cheese wheys)
Cellulose fibers (paper mills, board mills)
Glass fibers
Asbestos wastes
Vegetable wastes
Textile fiber wastes
Latex and rubber wastes
Polymeric wastes
Iron oxide scale (rolling-mill waters)
Paper-mill white water ("loading matter")
Paint-shop wastewater
Wool industry wash liquors

7

Separation and Recovery by Electrical Methods

Electromagnetic separation is based on the physical principle that unlike magnetic poles attract. When a particle is immersed in a magnetic field, a dipole field is induced in the particle and the field exerts a force on both ends of the particle. If the original field is constant, the forces on both ends of the particle will be equal but opposite and the particle will remain at rest. If the field has a gradient, then the force on one end of the particle will be larger than the force on the other end and the particle will be accelerated toward or away from the magnetic source depending on its magnetic properties. Ferromagnetic and paramagnetic materials are attracted and diamagnetic materials are repulsed.

This principle has been used to separate magnetic from nonmagnetic materials. Recent advances have been in the perfection of apparatus that can deliver high magnetic fields and gradients so that even weakly magnetic materials may be separated; this has opened the door to a multitude of applications which could not be carried out with weak fields in pollution control, that is, waste minimization as well as recovery of values.

Electromagnetic separation may be used to separate magnetic as well as nonmagnetic particles. With magnetic particles, the process consists of passing the material near a divergent magnetic field, attracting the particles to the separator magnets, and removing them. This has been effectively used in industry in:

- The removal of tramp iron from chemicals, foods, and lubricants. In this case, low-intensity magnets are sufficient.
- The beneficiation of kaolin to improve its grade and brightness. High-gradient magnetic separation (HGMS) has been very successful in this area and has replaced conventional techniques which are time consuming and require large land areas.
- The beneficiation of iron ores to improve their grade. Here again, the application of HGMS has been successful and its use in industry is expanding rapidly.
- The beneficiation of other minerals such as molybdenum, platinum, titanium, tungsten, tin, and uranium. A number of experimental electromagnetic separation systems have been used to successfully demonstrate the capabilities of HGMS.
- Solid waste material recovery. Traditionally, low-intensity magnets have been used to remove ferrous materials from municipal waste and refuse.
- The removal of magnetic materials from water effluents. This can have a twofold effect—the recovery of the materials, as well as the purification of the water.
- Coal desulfurization and ash removal. In this area, preliminary experimental results have indicated that coal may be purified with various degrees of success by the use of HGMS.
- Air pollution control to remove metallic particles from steel mill exhausts. Preliminary results are encouraging and future tests are planned.
- Medical applications. These include separation of blood cells, handling of encapsulated enzymes, purification of catalysts, and diagnostic and therapeutic applications.

With nonmagnetic particles, a magnetic seed is added together with a flocculent to the fluid to be treated. The flocculent binds the contaminants to the magnetic seed particles forming flocs that can be separated using conventional electromagnetic separators. This method has been effectively used to remove suspended solids, bacteria, viruses, and phosphates, and to modify the color and turbidity of river waters and discharges of sewage treatment plants. Results have been successful in demonstrating that magnetic separation in conjunction with other water treatment devices may be used to great advantage in water purification systems.

Another method utilizing electromagnetic separation uses a ferromagnetic fluid. The ferromagnetic fluid is placed in a divergent magnetic field which effectively varies its density as a function of the product of the field and its gradient. Hence, materials with varying density placed in the liquid will rise to different levels in the liquid depending on their density and can easily be separated. This method has been successfully used in the separation of various magnetic and nonmagnetic metals from refuse.

Other apparatus for electromagnetic separation include linear induction motors, eddy current, and magnetohydrodynamic separators that actually propel the materials to be separated out of the stream of the moving stock.

GENERAL PRINCIPLES

Traditionally, matter was considered to exist in three states—solid, liquid, and gas—but additional states of matter are now recognized.

1. *Solid State.* Solids have a definite shape and volume. Because of strong interactions between particles, their motion is limited.
2. *Liquid State.* Liquids have definite volume, but they assume the shape of the container. Particles are in clusters whose sizes are constantly changing because of weak attractive forces.
3. *Gaseous State.* Gases fill and assume the shape of their container. Particles are widely separated because they have such high velocities and attractions between them are very weak.
4. *The Plasma State.* A plasma is a fully ionized gas which has a sufficiently large number of charged particles to shield itself from externally applied electrostatic fields.
5. *The Superfluid State.* This state exists wherever thermal energy is limited enough to permit molecules to condense in momentum space.
6. *The Colloidal State.* This state exists when a solid or a liquid is so finely divided and dispersed that electrostatic forces overwhelm both gravitational and surface forces. The truly colloidal state exists at particle sizes below 0.1 micron. On the other hand, colloidal effects are important also at larger particle sizes. In practice, the term is loosely used when dealing with finely dispersed matter. Electromagnetic separation deals with particles in the colloidal state in the loose sense defined previously.

INDUCED MAGNETISM

Induced magnetism is observed whenever matter is placed in the magnetic field of a permanent magnet or electromagnet. The principal quantity used to describe the magnetic field is the magnetic induction, B. This vector quantity has for its sources all types of currents—true currents, J, and magnetization currents, J_m—as follows:

$$\text{curl } \vec{B} = \mu(\vec{J} + \vec{J}_m) \tag{7.1}$$

where μ is the permeability of free space. The currents, J, give rise to the magnetic intensity, $\vec{H}$:

$$\text{curl } \vec{H} = \vec{J} \tag{7.2}$$

and the magnetization currents which will be explained subsequently give rise to the magnetization, $\vec{M}$:

$$\text{curl } \vec{M} = \vec{J}_m \tag{7.3}$$

Each material is composed of atoms which have electrons in motion. The electrons form a current which circulates around the nucleus. Each atom can therefore be considered a magnetic dipole. In general, these dipoles are randomly oriented and the volume of material will have no net magnetic field.

When a volume of matter is placed in a magnetic field, some of these dipoles align themselves with the field. The number of dipoles that align themselves depends on the magnitude of the applied field as well as on the characteristics of the material, including its chemical composition, crystal structure, behavior with temperature, and previous history (some materials when placed in a magnetic field retain some magnetism after the field is removed). The field which results from this alignment is the magnetization, M. Most materials exhibit a linear relationship between the magnetization and the applied field. Logically, because of the continuity of the normal components of the field induction across a boundary, one would expect the magnetization to be defined proportional to B; however, long usage has associated M with H as follows:

$$\vec{M} = X_m \vec{H} \tag{7.4}$$

where $_m$ is a scalar called the susceptibility.

Substituting Equations 7.4, 7.3, and 7.2 in 7.1 we obtain:

$$\vec{B} = \mu(1 + X_m)\vec{H} \tag{7.5}$$

The quantity in the brackets in Equation 7.5 is known as the relative permeability, r, and:

$$\vec{B} = \mu\,\mu r\,\vec{H} = \mu\,\vec{H} \tag{7.6}$$

Equation 7.6 is the familiar matter equation relating B to H in a material with μ the permeability of the material. The susceptibility in Equation 7.4, X_m, is the volume susceptibility. In most handbooks, two other values of susceptibility may be listed and these are the mass susceptibility, X_m, mass and the molar susceptibility, X_m, molar given by:

$$\chi_m, \text{ mass} = X_m/p \tag{7.7}$$

$$\chi_m, \text{ molar} = X_m A p \tag{7.8}$$

where p and A are the density and molecular weight, respectively.

The susceptibility is a scalar quantity. In anisotropic materials, it is a tensor. In some materials, the alignment of the dipoles is such as to aid the original field. In that case, the susceptibility is positive and the material is paramagnetic. In other materials, the magnetization will oppose the original field and the susceptibility is negative. The material is then diamagnetic. It is possible for a material to be both paramagnetic and diamagnetic at the same time; for example, graphite specimens may be prepared to be paramagnetic in one direction and diamagnetic in another. In ferromagnetic materials, such as iron, nickel, cobalt and their alloys, the susceptibilities can have very large positive values and the relationship between $\vec{M}$ and $\vec{H}$ is nonlinear. In that case, the susceptibility is the slope of the $\vec{M}$-$\vec{H}$ curve.

Under some conditions, a magnetic field may be associated with a ferromagnetic material when some of the dipoles in the material are permanently aligned in a given direction. In this case a magnetization vector M_o is used to describe the field of the permanent magnet. Its source will be a magnetization current, J_{mo}, and can be shown as a current circulating on the periphery of the ferromagnetic material as shown in Figure 7.1.

Diamagnetism is based on the physical principle that the electronic currents in each atom are modified by the applied field so as to weaken the field. This effect is present in all substances except that in some materials it is masked by stronger paramagnetic or ferromagnetic effects. Materials which consist of atoms or ions with closed electron shells are usually diamagnetic. The diamagnetic susceptibility is given by:

$$\chi_m = \frac{-Ne^2\mu_0}{4m_e} \Sigma R_i^2 \cos^2\theta_i \tag{7.9}$$

where N is the number of molecules per unit volume; e, m_e and R_i, the charge, mass, and orbit radius of the electron, respectively; and $_i$, the angle between the magnetic field and the normal to the plane of electron rotation.

Paramagnetic substances, on the other hand, are characterized by atoms with a net dipole moment which tends to align itself with the field and hence

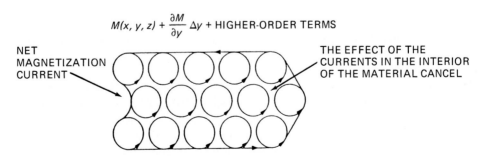

$$M(x, y, z) + \frac{\partial M}{\partial y} \Delta y + \text{HIGHER-ORDER TERMS}$$

NET MAGNETIZATION CURRENT

THE EFFECT OF THE CURRENTS IN THE INTERIOR OF THE MATERIAL CANCEL

Figure 7.1 Simplified picture of magnetic material consisting of atomic loop currents circulating in the same direction.

aid it. In this case, the susceptibility is given by:

$$\chi_m = \frac{Nm_o^2\mu_o}{3kT} \tag{7.10}$$

where m_o is the magnetic moment of the atom; k, the Boltzman's constant; and T, the absolute temperature.

For ferromagnetism to occur, the atoms which form magnetic dipoles must be very nearly aligned and the alignment maintained by the action of the molecular field. The Langevin function for ferromagnetic materials,

$$y = \frac{m_o\mu_o H_m}{kt} \tag{7.11}$$

which is normally small for paramagnetic materials, is larger than 0.3.

Ferromagnetism is exhibited by elements with unfilled inner electron shells, where the diameter ratio of the atom in the crystal lattice and the unfilled shell exceeds 1.5. Ferromagnetics change their linear dimension in the direction of the external field under magnetization. Those under positive striction are extended (permalloy); those with negative striction are contracted (nickel). The magnetic susceptibility of ferromagnetics is a complex function of field strength and temperature.

Ferromagnetism is exhibited by compounds with a specific crystal structurally formed moments. The ferromagnetic crystal represents a system of two lattices of mutually opposite magnetization, one inside the other; lack of full compensation of their magnetic moments causes some spontaneous magnetization. Ferrimagnetics are similar to ferromagnetics in that their susceptibility depends on the strength of the external magnetic field and temperature.

Antiferromagnetism usually appears only at the Neel temperature and persists above it. Antiferromagnetics have constant susceptibilities which do not depend on magnetic strength.

INTENSITY OF MAGNETIZATION

When various classes of substances are exposed to increasingly strong magnetic forces, three different kinds of magnetic behavior are observed (Figure 7.2).

1. *Ferromagnetic Group.* Strongly magnetic materials from this group are easily magnetized by a weak field. They respond rapidly, at first, to increasing magnetic fields. All the regions with paired north and south magnetic poles in the material become aligned. Rapidly, magnetization saturates the material; thereafter, the curve remains flat in spite of the increased magnetic field. The saturation level is then the magnetic field strength beyond which no further magnetization takes place. This level

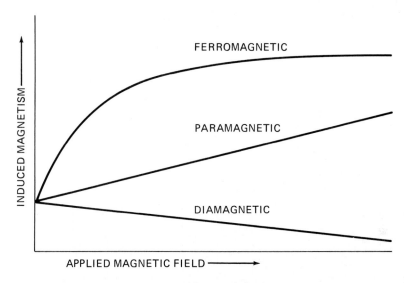

Figure 7.2 Magnetic behaviors.

depends heavily on the iron content in the material. Pure iron is saturated at a magnetization of some 220 electromagnetic units per cubic centimeter.

2. *Paramagnetic Group.* Weakly magnetic materials from this group show a lower magnetization response to applied fields; however, they rarely reach a saturation level. This means that weakly magnetic particles can be strongly magnetized if a sufficiently high magnetic field is used.

3. *Diamagnetic Group.* These materials show a slight but opposite magnetization in magnetic fields.

Magnetism and the Periodic Table

Figure 7.3 shows Mendeleev's periodic table with an indication of the ferromagnetic (3 elements) and paramagnetic (55 elements). Of these elements, 32 form compounds that are paramagnetic. Another 16 are paramagnetic in pure form but form diamagnetic compounds. The other seven elements become paramagnetic when one or more elements are present in a compound. Table 7.1 gives the magnetic susceptibility of the elements in their pure form.

Magnetic Susceptibility of Minerals

Minerals actually form a spectrum of chemical compounds and mechanical aggregates. Magnetic susceptibility of minerals represents the interaction of pure minerals and of the inclusions always present. Under these conditions, the concept of minerals is elusive and lacks a clear chemico-physical defini-

Figure 7.3 Magnetic characteristics of the elements.

TABLE 7.1   MAGNETIC SUSCEPTIBILITIES OF ELEMENTS

Element	Formula	Magnetic Susceptibility $\times 10^6$
Aluminum	$Al_{99.99}$	+0.61
Americium	Am	+4.00
Antimony	$Sb_{99.99}$	−0.75
Arsenic (amorphous crystalline)	As	−0.037
Barium	Ba	+0.15
Beryllium	Be	−1.00
Bismuth	Bi	−1.34
Boron	B	−0.62
Bromine	Br	−0.46
Cadmium	$Cd_{99.99}$	−0.176 (−0.310)
Calcium	Ca	+1.00
Carbon	C	+0.49
Cerium	Ce	+17.5
Cesium	Cs	+0.22
Chlorine	Cl	−0.57
Chromium	Cr	+3.5
Cobalt	Co	Ferromagnetic
Copper	Cu	−0.086
Dysprosium	Dy	+637.00
Erbium	Er	+265.0
Europium	Eu	+224.0
Gadolinium	Gd	+4800.0
Gallium	Ga	−0.33
Germanium	Ge	−0.105
Gold	Au	−0.15
Hafnium	Hf	+0.5
Helium	He	−0.47
Hydrogen	H	−1.99
Indium	In	−0.56
Iodine	I	−0.35
Iridium	Ir	+0.16
Iron	Fe	Ferromagnetic
Krypton	Kr	−0.344
Lanthanum	La	+0.85
Lead	Pb	−0.132
Lithium	Li	+0.50
Lutecium	Lu	0.00
Magnesium	Mg	+0.54
Manganese	Mn	+9.63
Mercury	Hg	−0.1667
Molybdenum	Mo	+0.93

(continued)

TABLE 7.1 MAGNETIC SUSCEPTIBILITIES OF ELEMENTS (*continued*)

Element	Formula	Magnetic Susceptibility $\times\ 10^6$
Neodymium	Nd	+39.01
Neon	Ne	−0.33
Nickel	Ni	Ferromagnetic
Niobium	Nb	+2.20
Nitrogen	N_2	−0.43
Osmium	Os	+0.064
Oxygen	O	+107.8
Palladium	Pd	+5.33
Phosphorus	P	−0.86 white
		−0.67 red
Platinum	Pt	+1.035
Plutonium	Pu	+2.52
Potassium	K	+0.53
Praseodymium	Pr	+35.6
Rhenium	Re	+0.363
Rhodium	Rh	+1.08
Rubidium	Rb	+0.198
Ruthenium	Ru	+0.427
Samarium	Sm	+12.1
Scandium	Sc	+7.0
Selenium	Se	−0.32
Silicon	Si	−0.13
Silver	Ag	−0.22
Sodium	Na	+0.70
Strontium	Sr	+1.05
Sulfur	S	−0.485
Tantalum	Ta	+0.849
Technetium	Tc	+2.7
Tellerium	Te	−0.31
Terbium	Tv	+917.0
Thallium	Tl	−0.249
Thorium	Th	+0.57
Thulium	Tm	+291.0
Tin	Sn	+0.026
Titanium	Ti	+3.19
Tungsten	W	+0.32
Uranium	U	+2.6
Vanadium	V	+5.00
Xenon	Xe	−0.334
Ytterbium	Yt	+1.44
Yttrium	Y	+2.15
Zinc	Zn	−0.175
Zirconium	Zr	+1.34

tion. Only the experimental approach can provide information on the magnetic properties of minerals. However, data collected by different laboratories under different conditions are often dissimilar and sometimes contradictory.

MAGNETIC SEPARATION—THEORETICAL CONSIDERATIONS

In general, the energy stored in the magnetic field in a volume, V, is given by:

$$E = \frac{1}{2} \int_V \vec{B} \cdot \vec{H} \, dv \tag{7.12}$$

The magnetic force, f, is the gradient of the stored energy or:

$$\vec{f} = \tfrac{1}{2} \text{grad} \int_V \vec{B} \cdot \vec{H} \, dv \tag{7.13}$$

For a paramagnetic or diamagnetic particle of volume, V, and magnetization, $\vec{M}$, the force acting on it when placed in a field is:

$$\vec{f} = \tfrac{1}{2} \text{grad} \int_V \vec{B} \cdot \vec{M} \, dv \tag{7.14}$$

V may be small, so that the magnetization and magnetic induction may be considered constant over the volume of integration and:

$$\vec{f} = \tfrac{1}{2} \text{grad} \, (\vec{B} \cdot \vec{M}) V$$
$$= \tfrac{1}{2} V (\vec{M} \cdot \text{grad}) \vec{B} \tag{7.15}$$

where we have assumed that $\vec{M}$ is constant and $\vec{M}$ and $\vec{f}_m$ in Equation 7.3 are collinear. Further simplification may be obtained by assuming that the magnetic intensity is a function of r (the radial coordinate) only. Then:

$$f_r = \tfrac{1}{2} V M \frac{dB}{dr} \tag{7.16}$$

It is obvious from Equation 7.16 that the force will be zero if the magnetic induction is constant; hence, for the force to exist, the components of the magnetic induction must have a gradient. Using Equation 7.14 we obtain:

$$f_r = \tfrac{1}{2} V \chi_m H \frac{dB}{dr} = \tfrac{1}{2} V (\chi_{mp} - \chi_{mm}) H \frac{dB}{dr} \tag{7.17}$$

where χ_{mp} and χ_{mm} are the susceptibilities of the particle and the medium in which the particle may be imbedded, respectively.

In a magnetic separator, the mixture to be separated is fed into the separator, as shown in Figure 7.4. The separator, usually a magnetic wire mesh, separates the magnetic particles (called *mags*) from the nonmagnetic

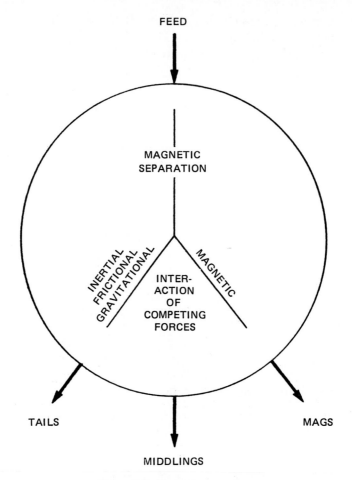

Figure 7.4   Magnetic separation.

particles (called *tails*). In addition, some particles which are less magnetic (called *middlings*) will be distributed between the mags and the tails. In the magnetic separator, there exist four types of forces:

- The tractive magnetic forces which will pull the magnetic material toward the wire mesh of the separator.
- The gravitational forces which act on the nonmagnetic materials so they fall away from the separator, but also act on the magnetic materials to dislodge them from the separator.
- The frictional forces which could remove the magnetic particles from the wire mesh.
- The attractive or repulsive interparticle forces.

These forces are obviously dependent on the nature of the feed as well as on the characteristics of the separation device, particularly the magnetic field and process rate.

In general, the geometry in the separator will be as shown in Figure 7.5. A strong magnetic field, H_o, will be applied to a ferromagnetic wire mesh. In Figure 7.5, a single filament of this mesh is shown. It is assumed that adjacent wires are sufficiently distant so as not to interact with each other. The field in the general vicinity of the wire will be strongly distorted by the induced dipole field in the wire. It is obvious that a magnetic particle will be

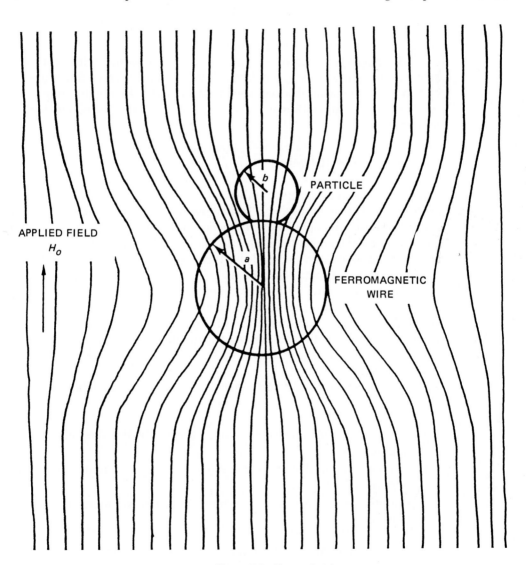

Figure 7.5 Competing forces.

attracted to the point of strongest field which is at the axis of symmetry. The field at the axis of symmetry is given by:

$$H = H_o \left(1 + \frac{a^2}{r^2} \right) \tag{7.18}$$

where H_o is the initial field and the second term in Equation 7.18 is the perturbation of this field due to the presence of the wire. For a spherical magnetic particle of radius b, Equation 7.17 becomes:

$$f_r = \frac{-4}{3} \mu\pi b^3 H_o^2 (\chi_{mp} - \chi_{mm}) \left(1 + \frac{a^2}{r^2} \right) \frac{a^2}{r^3} \tag{7.19}$$

where we have made use of Equation 7.6.

The magnetic force must compete in the separator with other forces acting on the particles traveling through the separator. These forces are those of gravity, hydrodynamic drag, friction, and inertia. In a high-gradient magnetic separator, gravitational and hydrodynamic drag forces are the most important.

MAGNETIC SEPARATOR EQUIPMENT

Magnetic separators are used either for concentration or purification. Concentration is the separation of a large amount of magnetic feed product. Purification is the removal of a small fraction of magnetic particles from a large amount of nonmagnetic feed material. From a practical point of view, separators are divided into wet and dry types. Figure 7.6 provides an overview classification of magnetic separators.

The most commonly used wet separators are permanent and electromagnetic drum separation systems, permanent and electromagnetic filters, and high-intensity separators. The wet magnetic drum separators are applied to concentrate finely divided, strongly magnetic material. They are essentially recovery units in dense-media plants and in the concentration of ferromagnetic iron ores.

Three basic types of drum separators are used for the following specific functions:

1. Concurrent drums for cobbing operations on material of minus 0.25 in. (Figure 7.7).
2. Counterrotation drums for roughing operations on minus 10-mesh ore (Figure 7.8).
3. Countercurrent drums for finishing operations on minus 65-mesh and finer ore.

Both permanent and electromagnetic types of drums are used; permanent magnets, however, represent the majority of the installations.

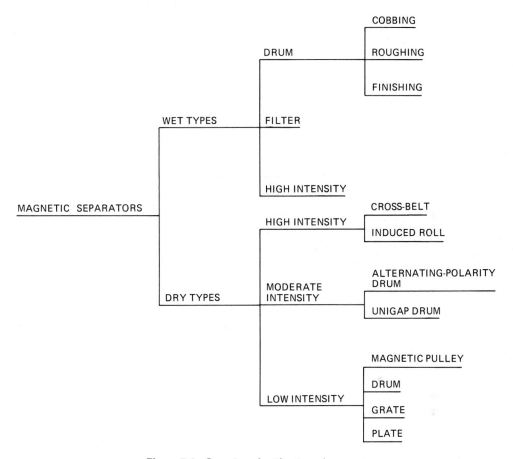

Figure 7.6 Overview classification of magnetic separators.

Most wet drums are 30 to 36 inches in diameter with magnet width to 10 feet. Cobbing and finishing drums are commonly double-drum or triple-drum separators. Feed volumes are typically:

cobbing separators   75/125 gal/min (ft of magnet width)
roughing separators   75/100 gal/min (ft of magnet width)
finishing separators   40–60 gal/min (ft of magnet width)

Operation and maintenance annual costs are on the order of 3 percent to 5 percent of initial cost.

Magnetic filters incorporate a filter element energized by an external permanent or electromagnet. The filters separate ferromagnetic particles from wet or dry materials of all kinds. There is a matrix for each material in the entire range from low-viscosity liquids to viscous fluids, and slurries through fine dry powders, coarse sands, and granular materials. A different matrix is selected in each case in order to provide maximum exposure of all particles

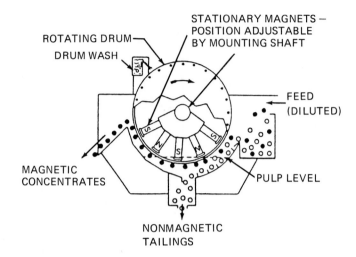

MAGNETICS

NONMAGNETICS

Figure 7.7   Concurrent wet-drum separator.

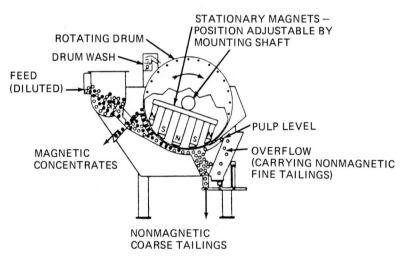

MAGNETICS

NONMAGNETICS

Figure 7.8   Counterrotation wet-drum separator.

to magnetic forces. For wet separation, the standard matrix is a stack of many grids. As the flow passes through the hundreds of uniform triangular openings of the first grid, it is divided into as many small streams. These streams are further split and resplit as they pass through the nonaligned openings of each successive grid.

Magnetic flux is concentrated along the edges of the triangles of the grids, which are fabricated of stainless steel ribbon of a high-magnetic permeability material. A strongly convergent magnetic field is maintained in the vicinity of each edge. The magnetic pull on ferromagnetic particles next to any grid edge is about 103 dynes/cm^3, or more than 10,000 times gravity. This magnetic force is sufficient to overcome a viscous drag of flowing liquid several thousand times the weight of a micron-sized particle.

Optimum separation occurs in the combined mechanical and magnetic combing of the fluid by the matrix. Particles suspended in a myriad of fractionated streams are repeatedly slowed by adhesion of the fluid to grid surfaces parallel to the flow. Low micron-sized particles are further slowed by molecular cohesion. Ferromagnetic particles slowed by these forces are arrested as they enter a converging magnetic field and are caught at the grid edge where magnetic flux is concentrated. Ferromagnetic particles carried through grid openings in midstream of any segment of the flow are repeatedly forced against the parallel walls of succeeding grids as each flow segment is divided and redirected again and again. For difficult separations the matrix contains several miles of magnetized edges within a small compass. Even the finest ferrous particles are slowed, arrested, and held. Yet the open structure of each grid permits full flow with low-pressure drop.

For dry materials the standard matrix is a stack of grids with inclined parallel vanes. The inclination of the vanes is reversed with each grid to guide the material through the matrix in a zigzag course. A variable eccentric weight vibrator imparts motion to the matrix to aid even distribution of grains and to ease the passage of the material. As it pours over a vane edge and falls on the face of the vane below, the material is turned over and broken up again and again. Vane faces retard the flow, and vane edges where magnetic flux is concentrated attract and hold the ferrous particles.

Cleaning is normally accomplished in a few seconds. The electromagnet is deenergized by a timer in automated systems to release the magnetic particles held on the grids. In wet processes, the matrix is flushed; in dry processes, the vibrator shakes out the contaminant. Automated cycling of both wet and dry process filters can be provided for continuous processing systems or very high flow rates.

Several commercially available include:

- Pipeline series—a completely enclosed flow.
- Underfeed series—designed for upward flow through the matrix and discharge of the cleaned material into a trough.
- Gravity series—designed for batch operations.

Special features including thermal insulation for handling hot liquids, explosion-proofing for hazardous locations, special matrices of various types, eccentric weight vibrators and automated systems are available.

Wet high-intensity separators are designed to remove fine weakly magnetic particles and minerals carried in liquid slurry. They extend the range

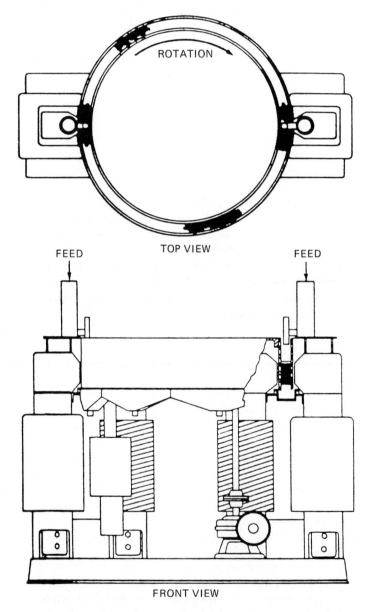

Figure 7.9 Continuous high-intensity wet magnetic separator.

of high-intensity magnetic separation into recovery of particles finer than 200-mesh size. The principle of operation of a continuous wet high-intensity separator is shown in Figure 7.9. The separator consists of an annular box rotating through one or more high-intensity magnetic fields. The annular box is filled with shaped ferromagnetic material inductively magnetized to a high magnetic field intensity and acting as a collecting medium for the weakly magnetic particles in the feed. The separator is capable of producing a field in excess of 20,000 gauss with a high field gradient between the induced poles. Non-magnetic particles pass directly through the annular ring, and the rotation of the annular box carries the magnetics out of the field where they are washed into a separate collection launder.

DRY MAGNETIC CONCENTRATORS AND PURIFIERS

Dry magnetic separators are usually classified on the basis of the intensity of the field used: low, moderate, or high.

High-Intensity Dry Magnetic Separators

Two types of high-intensity dry magnetic separators are recognized: the cross-belt and the induced roll. The crossbelt concentrates material by directly lifting the magnetics off the belt and discharging them to the side. The crossbelt concentration method is used to concentrate high-value mineral products. A typical 18-inch-wide crossbelt separator with two magnet poles will process up to two tons per hour of feed material.

The induced roll depicted in Figure 7.10 is used as a concentrator and a purifier in the chemical and mineral products industries. A common induced-roll 30-inch-wide threefold twin-type separator having 60 inches of total feed width will handle 3 to 7.5 tons of product per hour.

Moderate-Intensity Dry Magnetic Separators

The two types of moderate-intensity magnetic separators are the alternating-polarity drum separator (Figure 7.11) and the Unigap drum separator. The first separator has been used to concentrate magnetite ores in the 1.5-inch by 100-mesh size range. The magnetic field of this drum will hold a large load of magnetics while providing sufficient agitation of the magnetics during treatment to shake out impurities entrapped in the initial magnetic pickup. On coarser ores, capacities of 30 tons/hr per ft. of magnetic width have been obtained.

The Unigap drum is designed to produce a moderately high-intensity magnetic gap across the entire drum width at a single position on the drum circumference. This type of drum is useful in removing finely divided magnetic particles and is applied on material finer than 0.25 inch at feed rates of 3 tons/hr (ft of magnet width) or less.

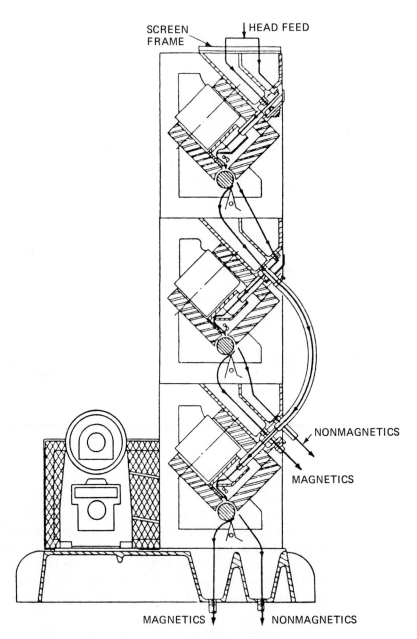

Figure 7.10 Operating principle of induced-role magnetic separator.

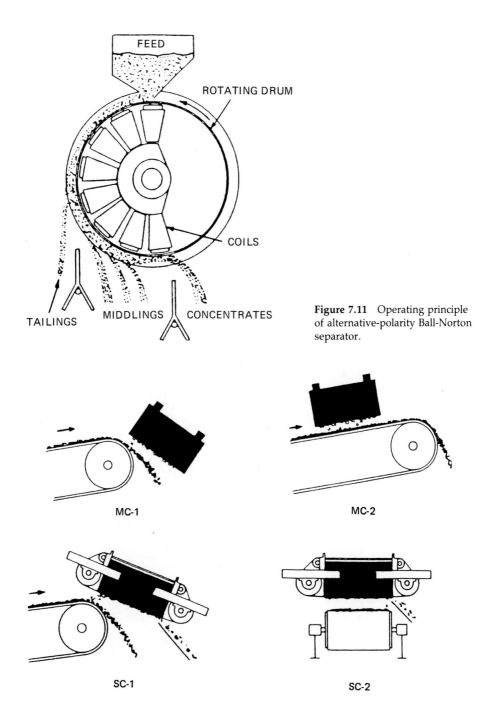

Figure 7.11 Operating principle of alternative-polarity Ball-Norton separator.

Figure 7.12 Magnetic pulleys.

Low-Intensity Dry Magnetic Separators

Low-intensity dry magnetic separators are applied in concentration and purification where the magnetic particles to be removed are large in size and the magnetic responsiveness of the magnetics is high. Magnetic pulleys (see Figure 7.12) and low-intensity tramp-iron drum separators are used in these applications, although the magnetics to be removed may be lower in responsiveness than tramp iron.

High-Speed Low-Intensity Dry-Drum Magnetic Separators

The high-speed alternating-polarity magnetic drum separator is designed to treat finely divided material, 100 mesh and finer, and to produce a high-grade magnetic concentrate product free of nonmagnetics. It has been used to con-

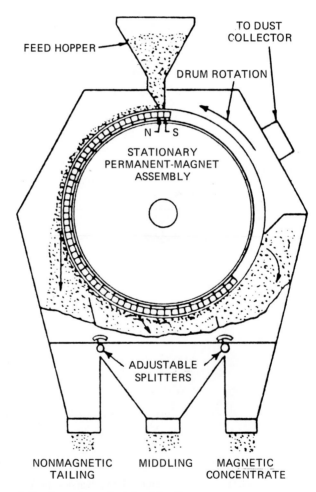

Figure 7.13 Operating principle of dry-fines permanent magnet separator.

centrate dry magnetite ores and fly ash. A 17-pole unit permits the application of this high-speed drum separator on particles as coarse as 1 inch. Figure 7.13 shows the general principle of operation.

HIGH-GRADIENT MAGNETIC SEPARATORS

Some of the modern high-intensity magnetic separators include:

The Jones Reciprocating Wet High-Intensity Magnetic Separators (WHIMS) consist of a box filled with grooved metal plates (Figure 7.14).

JONES PLATE BOX SEPARATOR

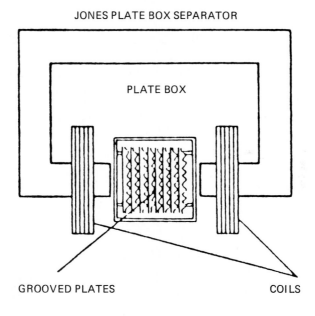

GROOVED PLATES   COILS

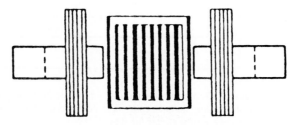

Figure 7.14 Jones Plate Box Separator. Schematic plan view (top) and elevation view showing steel yoke, coils, and plate box with grooved plates.

A magnetic field of up to 10 kilogauss is generated along the peaks of the grooves, and high gradient is created by the grooves. It is used extensively in the beneficiation of kaolin.

The rotating carousel version of the separator consists of a compartmented cylinder rotating vertically perpendicular to the magnetic field (Figure 7.15). Grooved plates are usually used, and fields of 20 kilogauss are achieved. This device is used for removing weakly magnetic particles as necessary in the kaolin industry.

JONES CAROUSEL SEPARATOR

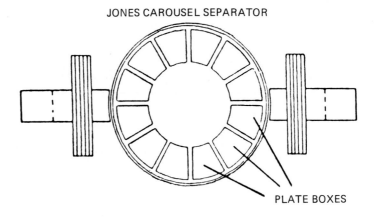

PLATE BOXES

ROTOR

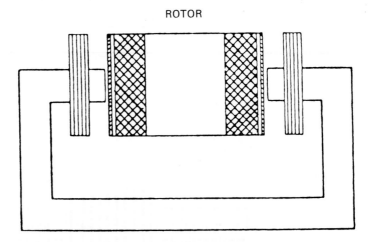

Figure 7.15 Carousel separator schematic plan view (top) showing compartmented rotor, magnet coils, and pole pieces. Elevation view (bottom) shows steel yoke return circuit for flux.

HYBRID SEPARATOR

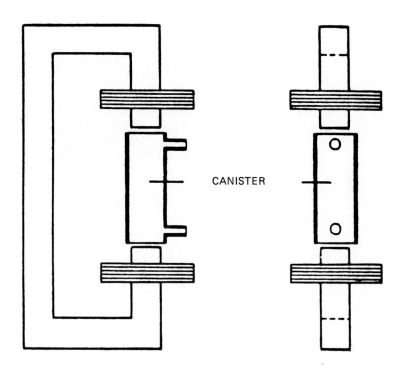

CANISTER

FIRST HIGH-INTENSITY/HIGH-GRADIENT SEPARATOR

Figure 7.16 Hybrid separator. Schematic elevation views showing coils, steel yoke magnet with canister in magnet gap.

The High-Intensity High-Gradient Hybrid Separator (Figure 7.16) combines the high-gradient and high-intensity field in a single machine.

High-Extraction Magnetic Filter (HEMF) machines using an iron-bound solenoid design, such as the machine shown in Figure 7.17, use a canister as long as 84 inches (210 cm).

Radial-Flow Filters, the HEMF described previously, is based on axial flow, that is the path of the slurry flow that is parallel or axial to the vertical field of the solenoid. A radial-flow canister can be substituted (Figure 7.18) where the slurry flow is introduced at the center of the canister and allowed to flow outwards.

LONG-COIL SEPARATOR

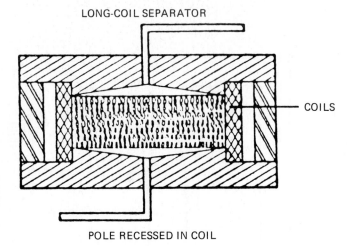

POLE RECESSED IN COIL

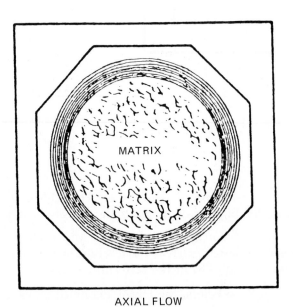

AXIAL FLOW

Figure 7.17 Long-coil separator. Schematic cross section in elevation (top) showing coil taller than canister and pole pieces fitting into the coil cavity. Fluid flow is parallel or axial to the magnetic field.

LONG-COIL SEPARATOR

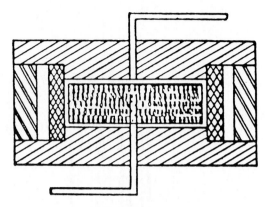

POLE RECESSED IN COIL

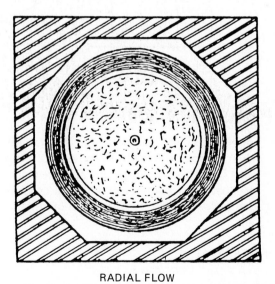

RADIAL FLOW

Figure 7.18 Long-coil separator. Schematic cross section showing fluid flow perpendicular or radial to the magnetic field.

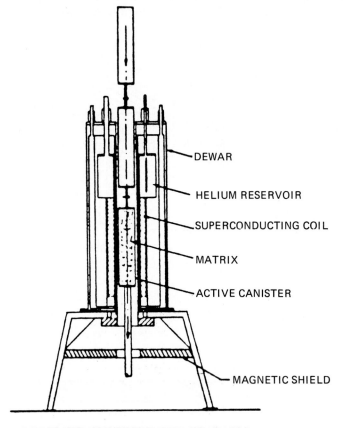

SUPERCONDUCTING MAGNETIC SEPARATOR
RECIPROCATING VERSION

Figure 7.19 Schematic of a separator.

Superconducting magnets offer several advantages and disadvantages when compared to the current state of the art. Experimental units have been fabricated (Figure 7.19) and are under test; however, at present, for most applications, conventional nonsuperconducting magnetic separators cost less to purchase and operate.

TRAMP-IRON REMOVAL

Tramp metal, usually defined as pieces of metal (typically iron) coarser than 0.125 inch, entering machinery can cause considerable damage to the equipment. It sometimes becomes white hot, starting destructive fires. Feed contaminated with sharp particles of tramp iron can cause intimal injuries and even death to livestock.

Using magnetic pulleys, grates, plates, or drums, tramp iron is easily separated from other nonmagnetic material and dropped out of the line of flow. In this way, a wide variety of materials is kept free of tramp irons.

Commonly used separators include:

- Magnetic Head Pulleys: Magnetic pulleys are used to remove the tramp iron mixed with materials moving on a belt conveyor. Both electromagnetic and permanent-magnet pulleys are used. Sizes and speeds range up to 66 inches in diameter and 500 feet/minute.
- Suspended Magnets: Magnets are installed flat over a conveyor head pulley.
- Magnetic Drums: Drums are used where space limitations prohibit the use of belt conveyors. They are usually used in combination with chutes or ducts in which the feed material flows.
- Plate Magnets: As feed flows through a 45°-angle chute, the tramp iron is trapped against a magnetized plate. Periodically the plate must be cleaned.
- Grate Magnets: A series of magnetized bars is used to collect fine iron as well as tramp iron. The feed is passed vertically through grates 1 inch in diameter and 2 inches apart. The grates must be periodically cleaned manually.
- Metal Detectors: In cases where the tramp metal is not of sufficient magnetic susceptibility to be removed by a magnet, an electronic metal detector is used to indicate its presence.

ELECTROCHEMICAL RECOVERY OF METALS

Processes using electricity to replace conventional processes using energy have various objectives, which may include energy savings with technical advantages in higher cost effectiveness. Also included may be simplicity of operation, ability to treat lower-grade ores, and the ability to recover by-product metals. At the same time environmental advantages are obtained, resulting in reduced emissions of gaseous species and reduced disposal of liquid effluents. These features make the processes attractive in the case of new mines, for they increase the recovery of primary and associated metals from mine wastes and lower-grade ores previously considered uneconomical. The major processes are listed in Table 7.2.

In the aluminum industry, major developments have included the introduction of the Alcoa process at the pilot plant stage and improvements to the conventional Bayer-Hall-Heroult process. The Alcoa process consists of forming aluminum chloride from alumina and electrolyzing it. The energy dissipated in the process is claimed to be 122.7×10^6 Btu/ton Al instead of 212×10^6 for the conventional process. Improvements to the conventional

TABLE 7.2 SUMMARY OF ELECTROWINNING PROCESS

Metal or Compound Recovered	Source Material	Problems of Conventional Process	Method
Aluminum	Alumina from bauxite or laterites	Bauxite supply, cost of electricity	Electrolysis of aluminum chloride
	Bauxite	Cell resistance fluorine emissions	Addition of lithium carbonate
	Bauxite	High anode-cathode distance	TiB_2 cathode plates
	Bauxite	Cell voltage high	Graphite block linings
	Bauxite	Cell temperature high	Lithium fluoride, low NaF/AlF_3 ratio
Antimony	Stibnite	Fossil fuel consumption, SO_2 emission	Electrooxidation with brine
Beryllium	Oxide	High cost, fossil fuel consumption	Fused-salt electrolysis
Beryllium-copper alloys	Beryllium oxide, copper rod	High power requirements, dust, and fume generation	Fused-salt electrolysis with dripping cathode
Cobalt	Sulfides	Fossil fuel consumption, SO_2 emission	Electrooxidation with brine
	Laterites	Low-grade ore	Reduction and ammoniacal leaching before electrolysis
Copper	Sulfides	Fossil fuel consumption, SO_2 emission	Ammonium-oxygen leaching
	Sulfides	Fossil fuel consumption, SO_2 emission	Ferric-chloride leaching
	Chalcopyrite	Fossil fuel consumption, SO_2 emission	Ferric-chloride leaching
	Mine waste dumps	Low-grade source material	Direct electrowinning from dilute acidic solutions
	Electrolytes	Low copper content	Improvements of current density
	Chalcopyrite	Fossil fuel consumption, SO_2 emission, insoluble ferrites, sulfate disposal	Two-stage oxidation-reduction roasting leach electrowinning
	Chloride solutions	Pollution when reducing	Copper chloride electrolysis

Metal	Material	Problem	Process
Gold	Ores containing clay	Conventional process impractical because of ore components	Cyanide heap leaching
	Slimy ores	Conventional process impractical because of ore components	Cyanide leaching followed by carbon extraction from pulp (carbon-in-pump)
	Carbonaceous ores	Organic compounds interfere with conventional process	Electrooxidation with brine
Iron	Iron-titanium oxides	—	Electrosmelting
Lead	Sulfide ores	Fossil fuel consumption, SO_2 emission	Electrooxidation in diaphragm cell
	Sulfide ores	Fossil fuel consumption, SO_2 emission	Leaching with $FeCl_3$-NaCl
	Complex lead-zinc-copper ores	Fossil fuel consumption, SO_2 emission	Chlorine-oxygen leaching and fused-salt electrolysis
	Lead chloride	—	Fused-salt electrolysis
Mercury	Cinnabar	Fossil fuel consumption, SO_2 emission	Electrooxidation with brine
Molybdenum	Carbide	—	Electrolysis
	Technical grade molysulfide	—	Aluminothermic reduction and fused-salt electrorefining
	MoS_2-C	—	Fused-salt electrolysis
	Molybdenum and rhenium sulfides (molybdenite)	Fossil fuel consumption, SO_2 emission	Electrooxidation with brine
Nickel	Laterites	Low-grade ore	Ammoniacal leaching and electrolysis in diaphragm cell
	Ferronickel	—	Solvent extractions and electrodeposition
Rare earths	Oxides	Insufficient purity, chlorine emission	Electrolysis in fluorides melt
	Bastnasite	Chlorine emission	Leaching, roasting, drying, electrolysis in fluorides melt
	$LiYF_3$, magnesium	Low melting point of magnesium	Floating liquid cathode electrolytic cell
Silver	Low-grade ores, mine wastes	Low concentration	Heap leaching and precipitation with sodium sulfide

(continued)

TABLE 7.2 SUMMARY OF ELECTROWINNING PROCESS (*continued*)

Metal or Compound Recovered	Source Material	Problems of Conventional Process	Method
	Silver sulfides	Fossil fuel consumption, SO_2 emission	Electrooxidation with brine
Tungsten	Wolframite	Fossil fuel consumption	Separation of halide tungstate melt and electrolysis
	Scheelite	Accumulation of lime in electrolyte	HCl digestion and electrolysis of oxide
Uranium	Ore	Yellow-cake preparation step to be avoided	Electrolysis of uranyl chloride
Vanadium	Carbide	Purity	Electrolysis of vanadium dichloride
Zinc	Sulfides	Fossil fuel consumption, SO_2 emission	Chlorine-oxygen leaching, fused-salt electrolysis
	Ore	Zinc ferrite losses	Jarosite precipitation
Zirconium	Chlorides	Cell design	Second chamber for tetrachloride addition
Zirconium diboride	Oxide or zircon (silicate)	—	Fused-salt electrolysis

process include introduction of lithium carbonate into the electrolyte to lower its internal resistance, titanium diboride cathode plates to reduce anode-cathode distance, graphite cell linings to reduce cell voltage, and a modified electrolyte including lithium fluoride to lower cell operating temperature to 900°C.

Antimony often occurs in nature in the form of stibnite, which is a sulfide. The electrooxidation process for sulfides developed by the Bureau of Mines for application to a whole series of metals can be used with a stibnite. The ore, crushed and finely ground, is mined with a brine solution and electrolyzed. An oxidizing species, OCl in hypochlorite HOCl, is produced from the brine and oxidizes the metal sulfide to a metal oxide, sulfate, or chlorocomplex. With stibnite, the oxide Sb205 is produced. Subsequent recovery of the metal from the oxides or sulfates follows classical paths such as reduction, or fused-salt electrolysis for the oxides, and precipitation with an active metal for the sulfates and chlorocomplexes.

For beryllium, the Bureau of Mines has developed a special electrowinning cell for molten-salt electrolysis of the oxide at 780°C along with procedures to transform the electrodeposited metal into either powder or spheres. It has also developed a cell permitting the preparation of copper-beryllium alloys with up to 4 percent beryllium. This is accomplished by the use of a new "dipping cathode" in the cells where the low-melting product alloy is made to drip from the cathode into a collection crucible.

Cobalt appears mostly in sulfides and low-grade laterites. It can be recovered from sulfides by the electrooxidation process described in the antimony section. Cobalt can be prepared from laterites, containing nickel with the cobalt, by a new Bureau of Mines process using pyrite and carbon monoxide reduction followed by ammoniacal leaching and solvent extraction performed before the electrowinning of the nickel.

Copper metallurgy is an area in which industry is beginning to introduce most of the newly developed processes. For example, Kennecott has begun to use its direct electrowinning process, where innovative cathodes are introduced, consisting of thinly packed beds of fine coke which are spaced between the anodes. Other efforts are directed at increasing the current density in the cells by means such as periodic current reversal, electrolyte motion enhancers, air agitation, and high-frequency sound. The Bureau of Mines has developed a two-stage process for treatment of chalcopyrite ($CuFeS_2$) consisting first in an oxidation of the sulfides to copper-iron oxides and second in a reduction to copper metal and iron oxide. The Bureau of Mines has also developed cells and methods to electrolyze the copper chlorides obtained when its electrooxidation process is applied to sulfides or chalcopyrite or other ores.

Various processes have been historically explored for gold. In the cyanide heap leach process, the ore is first leached with a lime cyanide solution; then the solution is passed through activated carbon. Next, the carbon is stripped of its gold and, finally, the stripping solution is electrolyzed. The carbon-in-pulp process differs in that it involves a slurry which is vigorously

agitated and aerated during the leaching process instead of simple heap leaching. Electrooxidation, as already described for antimony, has been successfully applied by the Bureau of Mines to carbonaceous ores where the presence
of organic compounds interferes with cyanide extraction.

Iron oxides are not currently electrolyzed in the United States nor is
research done on them.

For lead, the efforts have been enthusiastic and all concentrate on transforming sulfides into chlorides. One process uses electolytic oxidation in diaphragm cells; the second utilizes leaching with a mixture of iron chloride and
sodium chloride; the third employs direct chlorine-oxygen leaching. The lead
and zinc present are then electrowon from the chlorides by fused-salt electrolysis.

Mercury is processed from cinnabar ores (sulfides) to which the electrooxidation process was again found to be applicable. Extraction of the electrowon metal is then quite naturally done by amalgamation with an active
metal and sublimation of the amalgam.

Molybdenum has been the subject of a whole series of investigations.
The carbide could be electrolyzed. Sulfides have been treated by various methods including aluminothermic reduction followed by electrorefining, fused-
salt electrolysis of pellets of sulfide compacted with graphite, and electrooxidation by the Bureau of Mines process for which a new bipolar flow-through
cell was invented.

Nickel has focused on lateritic ores (very low-grade deposits in terrains
that have been transformed by rains into tropical or semitropical climates)
and on separating nickel from the easily manufactured ferronickel alloy. For
laterites, ammoniacal leaching was used (as for cobalt) and special diaphragm
cells were developed for the subsequent electrolysis. For ferronickel, electrolytic purification by successive solvent extraction procedures was developed.
Chlorite was used instead of sulfate, and a specially selected amine was used
to separate iron and other impurity metals from the nickel-bearing solution.

The Bureau of Mines' efforts on rare earths have been continuous. Efforts
have been directed at recovery of high-purity alloys of high-melting rare
earths in new fluoride baths with cells operating at temperatures as high as
1,700°C and with the possibility of tapping the liquid electrowon product.
Misch metal was electrowon from bastnasite (mixture of oxides and fluorides
of rare earths) after leaching and washing to remove carbonates and baryte
present in the ore, using a new cell in which the liquid product could again
be tapped. Preparation of magnesium-yttrium alloys necessitated development of a special cell with a liquid product cathode.

Silver can be recovered as a by-product from silver-gold ores by means
of a process developed by the Bureau of Mines. The ore is heap leached with
a lime-cyanide solution and silver values are precipitated with sodium sulfide.
The silver can then be separated by simple filtration, while the gold is separated from the solution by carbon absorption prior to carbon stripping and
electrowinning. Primary silver ores (sulfides) have also been investigated and

the Bureau of Mines' electrooxidation process described earlier has been tried with success to recover this precious metal.

Tungsten generally comes in two forms: wolframite (tungstate of iron and manganese) and scheelite (tungstate of calcium). Wolframite was mixed with various sodium salts, including chloride, heated to 1,080°C and a fluid halide tungstate was easily separated from the rest of the melt. This halide tungstate could then be electrolyzed in an experimental cell. The scheelite route is different. After leaching with hydrochloric acid and calcining the tungstic acid found, tungsten oxide is obtained, which is subsequently electrolyzed.

Uranium can now be obtained directly as tetrafluoride with a process which avoids the yellow-cake stage. Developed by a Japanese firm (Asahi), the process utilizes an amine to extract uranium as a sulfate complex from the ore-leaching solution. Uranyl chloride is then fabricated with hypochloric acid and processed in a diaphragm electrolytic cell to give uranium tetrachloride. Reaction with hydrofluoric acid gives the required tetrafluoride.

The electrolysis of vanadium carbide has been achieved in a special helium atmosphere cell utilizing fused chlorides as a bath. Vanadium of 98 percent to 99 percent purity was obtained and then electrorefined, attaining the 99.96 percent purity level.

A procedure for producing special high-grade zinc from sulfide concentrates employs chlorine-oxygen leaching followed by fused-salt electrolysis of the chlorides formed in a lithium and potassium chloride bath. Zinc chloride is evaporated to obtain anhydrous chloride for cell feed. Lead is often found with the zinc, and can also be won by the electrolysis of its chloride. Copper, cadmium, and silver, if present, can be extracted by conventional means from the zinc-bearing filtrate. Work has also been done on existing processes. One (the jarosite process) consists of selectively precipitating deleterious iron with sodium to prepare an iron-depleted solution suitable for electrolysis and zinc recovery. Another improvement is preleaching with dilute sulfuric acid to remove magnesium. In addition, automation of cathode handling and stripping, and new lead oxide-coated titanium anodes, are gradually being introduced into the zinc industry.

Efforts in electrochemical zirconium recovery have resulted in the electrolysis of zirconium tetrachloride in an improved cell featuring a second chamber for tetrachloride addition and transfer. Zirconium diboride, useful in industrial applications, can be prepared directly from the oxide and from zircon (the silicate) in a high-temperature cell featuring a molten salt bath of sodium diboride and sodium and aluminum fluorides.

RECOVERY OF BYPRODUCT METALS

Ores do not occur in nature in a pure form; rather, they usually contain a mixture of several metals. The metal that represents the major percentage of the ore is usually the one for which the ore is mined. Sometimes ore is mined

TABLE 7.3 ELECTROWINNING PROCESSES BY SOURCE MATERIAL

Source Material	Process	Metal or Product
Sulfide	Electrooxidation with brine	Antimony, cobalt, mercury, molybdenum, rhenium, silver
	Chlorine-oxygen leaching with fused-salt electrolysis	Zinc
	Ammonia-oxygen leaching (Arbiter)	Copper
	Ferric chloride leaching (Clear)	Copper
	Electrooxidation in diaphragm cells	Lead
	Leaching with $FeCl_3$-$NaCl$	Lead
	Aluminothermic reduction and fused-salt electrorefining	Molybdenum
Oxide	Fused-salt electrolysis	Molybdenum
	Fused-salt electrolysis	Beryllium
	Electrosmelting	Iron titanium
	Electrolysis in fluorides melt	Rare earths
	Fused-salt electrolysis	Titanium-, zirconium-, hafnium-diborides
Carbonate	HCl digestion and electrolysis of oxide	Tungsten
Carbide	Electrolysis with fused salts	Molybdenum
	Electrolysis of dichlorides	Vanadium
Chloride	Chloride solution electrolysis	Copper
	Twin-cell electrolysis	Zirconium
	Fused-salt electrolysis	Lead
Silicate	Fused-salt electrolysis	Zirconium diboride
Bauxite	Alcoa	Aluminum

Laterites	Reduction ammoniacal leaching	Aluminum
	Ammoniacal leaching and electrolysis	Cobalt
	in diaphragm cell	Nickel
Chalcopyrite (CuFeS$_2$)	Ferric chloride leaching (Cymet)	Copper
	Two-stage oxidation reduction roasting	Copper
	leach electrowinning	
Complex lead-zinc-copper ore	Chlorine-oxygen leaching with fused-salt electrolysis	Lead, zinc
Ores containing clay	Cyanide heap leaching	Gold
Slimy ores	Carbon-in-pulp	Gold
Carbonaceous ores	Electrooxidation with brine	Gold
Ferronickel	Solvent extraction and electrodeposition	Nickel
Mine waste dumps	Direct electrowinning from dilute acidic solutions (Kennecott)	Copper
	Heap leaching and precipitation with sodium sulfide	Silver
Bastnasite concentrate (mixture of oxides with some fluorides of rare earths)	Leaching, roasting, drying electrolysis in fluoride melts	Misch metal (cerium, lanthanum, neodymium, praseodymium, and so on)
Wolframite (tungsten, iron, manganese composite oxide)	Separation of halide-tungstate melt and electrolysis	Tungsten
Uranium ores	Electrolysis of uranyl chloride	Uranium tetrafluoride
Lithium-yttrium fluorides with magnesium metal	Flating liquid cathode electrolytic cell	Yttrium-magnesium alloy
Zinc-iron ores	Jarosite precipitation	Zinc, cadmium, silver, lead

for two or more metals at a time; typical groups of metals that may appear together are listed in Table 7.3. The introduction of electrolytic steps at the end of the main recovery process can achieve recovery of one or several additional metals that would otherwise be lost in the tailings or wastewaters. This is economicallly advantageous in cases where the secondary metals are precious, such as gold, silver and platinum, or rare and expensive, such as the rare earths.

ADVANTAGES

Besides incentives, such as process simplification and by-product recovery which are mainly geological in nature, there may be others which are purely technical. Conventional milling operations involve many steps—from crushing the ore to drying the concentrate, including grinding, flotation, upgrading, and dewatering—before the metal recovery operation proper can start. From this point, either chemical (wet) processes or high-temperature (calcination) processes take over, involving many stages before the metal can be recovered. Introduction of electrolytic techniques, some of which can be performed at the mill, can reduce the number of steps, avoid complicated chemical procedures, reduce fossil fuel consumption, and make transportation of voluminous, heavy ore concentrates unnecessary.

Economic advantages are usually tied to technical progress in the form of a reduction in capital, operation, and maintenance costs. They can also derive from savings in energy and materials, especially if the extraction efficiency of the metal from the ore is improved, in other words, if the tails assay of the important metals is lowered. An increase in the quantity and value of by-products, heretofore not recoverable by previous methods, is also a bonus.

Important advantages expected from the introduction of electrochemical methods for metal recovery from minerals are also found in the environmental area. The avoidance of thermal processes involving calcination of the ore, decomposition of the same at high temperature, emission of flue gases from the fuels used, and of chemical gaseous species and particulates from the minerals is invaluable.

Wastewaters and other residues or tailings in aqueous solutions are amenable to recovery of the included metals by electrolysis. In that case, the material dispersed to the environment or going back to the water table can be purified by the electrochemical recovery process so that pollution standards can be met.

VARIOUS TYPES OF PROCESS INNOVATIONS

Due to the great variety of source materials and the great difference in chemical properties of the metals to be recovered, each problem must be studied individually. In some cases, completely new types of electrolytic cells and procedures must be developed to perform the electrolysis and recover the metal in pure form as a deposit on the cathode. The structure of the cell, the chemistry of the electrolyte, and the operating characteristics had to be experimentally developed. To achieve satisfactory electrolysis, a variety of configurations has been utilized including flow-through cells, diaphgram cells, high-temperature cells, cells with continuous electrolyte feed and/or tapping of the product, special anodes for high-current density, dripping or liquid cathodes when the melting temperature of the recovered metal was below that of the bath, phase separation at high temperature, and the addition of a number of special chemicals to the feed salts to facilitate their melting and lower their sensitivity.

However, the electrolysis itself was not always the major point of emphasis of the research and development efforts. In many instances, the major difficulties are encountered in the process of transforming the ore or mineral or source material available into a feed suitable for subsequent electrolysis. Consequently, chemical and physical methods have to be found to make the electrolysis possible which include processes such as flotation; phase separation; leaching with acids, chemical species or compounds (such as chlorides), or gases (oxygen); solvent extraction; oxidations (sometimes performed in the electrolytic cell itself) reduction; and, occasionally, classical fossil fuel-consuming methods such as roasting, drying, and smelting.

A third type of process is applicable after electrolysis has taken place. In some cases it is necessary to extract or isolate the desired metal from the products obtained after electrolysis, either because the desired metal was mixed with other species or because it was recovered in the form of a chemical compound that had to be broken down (for example, chlorides in a pregnant solution). Such processes include precipitation by contact with an active metal, solvent extraction, ion-exchange techniques, carbon absorption and, of course, recycling of the electrolyte.

When reviewing the processes for the purpose of choosing a process for a new ore or when comparing processes among themselves, it appears that initially an appropriate basis for grouping is the type of source material. This review effort reveals typical families of ores or minerals which can all be treated by similar or nearly equivalent methods to obtain a whole series of metals. Major typical families include, for instance, the sulfides, many of which are amenable to treatment by the Bureau of Mines' method of electrooxidation with brine; or the oxides, which are usually electrolyzed as fused salts in electrolytes to which various fluorides have been added to lower cell voltage and bath sensitivity. Some source materials are technical-grade chemical compounds industrially produced, such as carbides, chlorides, alloys; others are present in mine dumps, tailings, or recycled electrolytes. Another category of source materials is composite ores bearing several metals like chalcopyrite (copper iron sulfide), complex lead-zinc-copper ores, bastnasite (mixture of rare-earth oxides), and others, from which it is usually desired to recover the various components. The process is accordingly more complex

TABLE 7.4 ELECTROWINNING PROCESSES ACCORDING TO MENDELEEV'S GROUPS OF METALS

Metal or Compound Material	Source Material	Problems of Conventional Process	Method
Group I—Copper, Silver, Gold			
Copper	Sulfides	Fossil fuel consumption, SO_2 emission	Ammonia-oxygen leaching
	Sulfides	Fossil fuel consumption, SO_2 emission	Ferric-chloride leaching
	Chalcopyrite	Fossil fuel consumption, SO_2 emission	Ferric-chloride leaching
	Mine waste dumps	Low-grade source material	Direct electrowinning from dilute acidic solutions
	Electrolytes	Low copper content	Improvements of current density
	Chalcopyrite	Fossil fuel consumption, SO_2 emission, insoluble ferrites, sulfate disposal	Two-stage oxidation reduction, roasting, leach, electrowinning
	Chloride solutions	Pollution when reducing	Copper chloride electrolysis
Silver	Low-grade ores, mine wastes	Low concentration	Heap leaching and precipitation with sodium sulfide
	Silver sulfides	Fossil fuel consumption, SO_2 emission	Electrooxidation with brine
Gold	Ores containing clay	Conventional process impractical because of ore components	Cyanide heap leaching
	Slimy ores	Conventional process impractical because of ore components	Cyanide leaching followed by carbon extraction from pulp (carbon-in-pulp)
	Carbonaceous ores	Organic compounds interfere with conventional process	Electrooxidation with brine
Group II—Beryllium, Zinc, Mercury			
Beryllium	Oxide	High cost, fossil fuel consumption	Fused-salt electrolysis
Beryllium-copper alloys	Beryllium oxide, copper rod	High power requirements, dust and fume generation	Fused-salt electrolysis with dripping cathode

Zinc	Sulfides	Fossil fuel consumption, SO_2 emission	Chlorine-oxygen leaching, fused-salt electrolysis
		Zinc ferrite losses	Jarosite precipitation
Mercury	Ore		
	Cinnabar	Fossil fuel consumption, SO_2 emission	Electrooxidation with brine

Group III—Aluminum, Rare Earths

Aluminum	Alumina from bauxite or laterites	Bauxite supply, cost of electricity	Electrolysis of aluminum chloride
	Bauxite	Cell resistance, fluorine emissions	Addition of lithium carbonate
	Bauxite	High anode cathode distance	TiB_2 cathode plates
	Bauxite	Cell voltage high	Graphite block linings
	Bauxite	Cell temperature high	Lithium fluoride, low NaF/AlF_3 ratio
Rare earths	Oxides	Insufficient purity, chlorine emission	Electrolysis in fluorides melt
	Bastnasite	Chlorine emission	Leaching, roasting, drying, electrolysis in fluorides melt
	$LiYF_3$, magnesium	Low melting point of magnesium	Floating liquid cathode electrolytic cell

Group IV—Zirconium, Lead

Zirconium	Chlorides	Cell design	Second chamber for tetrachloride addition
Zirconium diboride	Oxide or zircon (silicate)	—	Fused-salt electrolysis
Lead	Sulfide ores	Fossil fuel consumption, SO_2 emission	Electrooxidation in diaphragm cell
	Sulfide ores	Fossil fuel consumption, SO_2 emission	Leaching with $FeCl_3$-NaCl
	Complex lead-zinc-copper ores	Fossil fuel consumption, SO_2 emission	Chlorine-oxygen leaching and fused-salt electrolysis
	Lead chloride	—	Fused-salt electrolysis

(continued)

TABLE 7.4 ELECTROWINNING PROCESSES ACCORDING TO MENDELEEV'S GROUPS OF METALS (*continued*)

Metal or Compound Material	Source Material	Problems of Conventional Process	Method
Group V—Vanadium, Antimony			
Vanadium	Carbide	Purity	Electrolysis of vanadium dichloride
Antimony	Sulfide (stibnite)	Fossil fuel consumption, SO_2 emission	Electrooxidation with brine
Group VI—Molybdenum, Tungsten, Uranium			
Molybdenum	Carbide	—	Electrolysis
	Technical grade molysulfide	—	Aluminothermic reduction and fused-salt electrorefining
	MoS_2-C	—	Fused-salt electrolysis
	Molybdenum and rhenium sulfides (molybdenite)	Fossil fuel consumption, SO_2 emission	Electrooxidation with brine
Tungsten	Wolframite	Fossil fuel consumption	Separation of halide-tungstate melt and electrolysis
	Scheelite	Accumulation of lime in electrolyte	HCl digestion and electrolysis of oxide
Uranium	Ore	Yellow-cake preparation step to be avoided	Electrolysis of uranyl chloride
Group VIII—Iron, Cobalt, Nickel			
Iron	Iron-titanium oxides	—	Electrosmelting
Cobalt	Sulfides	Fossil fuel consumption, SO_2 emission	Electrooxidation with brine
	Laterites	Low-grade ore	Reduction and ammoniacal leaching before electrolysis
Nickel	Laterites	Low-grade ore	Ammoniacal leaching and electrolysis in diaphragm cell
	Ferronickel	—	Solvent extractions and electrodeposition

and includes operations for separating and recovering the various metals desired.

A categorization according to the source material has been made for the major processes described in this report; the categorization is presented in Table 7.3. It identifies for each kind of ore or mineral the processes utilized and the metals recovered with each of them.

A second appropriate basis for a comparison of the processes is to use the metal groups as defined in Mendeleev's classification. Each group corresponds to a column in Mendeleev's table and includes chemical species having similar chemical properties. It can then be inferred that if one process has been applicable for one metal of the group when found in a particular ore, it will then, in principle, be applicable to another metal of the same group if this metal is found in an ore of similar chemical composition. A table of this kind might thus be very useful for identifying the appropriate process for new ores or minerals, as it gives an immediate review of all the processes already developed for various forms of source material of metals in the group.

For ease of identification, the group numbers (number of column in Mendeleev's table) are indicated for the elements under consideration; also see Table 7.4.

Aluminum	III	Cobalt	VIII
Antimony	V	Copper	I
Beryllium	II	Gold	I
Iron	VIII	Silver	I
Lead	IV	Tungsten	VI
Mercury	II	Uranium	VI
Molybdenum	VI	Vanadium	V
Nickel	VIII	Zinc	II
Rare Earths	III	Zirconium	IV

The elements for which electrochemical recovery methods have been indicated are distributed among the groups in Mendeleev's table as follows:

Group I	Copper Silver Gold	Group V	Vanadium Antimony
Group II	Beryllium Zinc Mercury	Group VI	Molybdenum Tungsten Uranium
Group III	Aluminum Rare Earths	Group VII	Not represented
Group IV	Zirconium Lead	Group VIII	Iron Cobalt Nickel

  8

Carbon Adsorption

Adsorption is the selective collection and concentration onto solid surfaces of particular types of molecules contained in a liquid or a gas. By this unit operation, gases or liquids of mixed systems, even at extremely small concentrations, can be selectively captured and removed from gaseous or liquid streams using a wide variety of specific materials known as adsorbents. The material which is adsorbed onto the adsorbent is called the adsorbate. Two mechanisms involved are chemical adsorption and physical adsorption. We focus specifically on carbon adsorption.

When gaseous or liquid molecules reach the surface of an adsorbent and remain without any chemical reaction, the phenomenon is called physical adsorption, or physisorption. The mechanism of physisorption may be intermolecular electrostatic or van der Waals' forces or may depend on the physical configuration of the adsorbent such as the pore structure of activated carbon. Physical adsorbents typically have large surface areas. The properties of the material being adsorbed (molecular size, boiling point, molecular weight, and polarity) and the properties of the surface of the adsorbent (polarity, pore size, and spacing) both serve to determine the quality of adsorption. There are also the following parameters which can be used to improve physical adsorption:

- Increase the adsorbate concentration.
- Increase the adsorbate area.

- Select best adsorbent for the specific gas system.
- Remove contaminants before adsorption.
- Reduce the adsorption temperature.
- Increase the adsorption contact time.
- Frequent replace or regenerate the adsorbent.

Physical adsorption units may either be regenerable types or use disposable adsorbents. Regeneration of physical adsorbents is accomplished via any combination of three mechanisms, namely temperature, pressure, and concentration swings. Freshly regenerated adsorbents theoretically remove 100 percent of the contaminants and, at the other extreme, significant quantities of contaminants begin to escape at the breakthrough point. Physical adsorption systems may either consist of two beds (adsorption, desorption) or three beds (adsorption, desorption, cooling).

When gaseous or liquid molecules adhere to the surface of the adsorbent by means of a chemical reaction and the formation of chemical bonds, the phenomenon is called chemical adsorption or chemisorption. Heat releases of 10–100 kcal/g-mol are typical for chemisorption, much higher than the heat release for physisorption. With chemical adsorption, which is far less common than physisorption, regeneration is often either difficult or impossible. Chemisorption usually occurs only at temperatures greater than 200°C when the activation energy is available to make or break chemical bonds.

Activated carbon is the most widely used adsorbent today. It is usually categorized as a physical adsorbent and also as a nonpolar adsorbent. It can be produced from a wide variety of carbonaceous materials and provides an extremely high internal surface area within its intricate network of pores. A total surface area range of 450 to 1,800m²/gram has been estimated. Only a portion of that area is available for adsorption in pores of the proper size. For organic solvent adsorption, carbonaceous material is capable of removing at least 85 percent of the emissions. Activated carbon comes in three general types: granular or natural grains, pellets, and powders. The natural grains which are hard and dense are most appropriate for gaseous-phase adsorption applications while other types, liquid-phase adsorbents, are commonly used to decolorize or purify liquids and solutions. Generally, liquid-phase carbons have about the same surface areas as gas-adsorbing carbons, but have larger total pore volumes. Liquid-phase carbons are generally either powdered or granular, the former mixed and later filtered from the liquid, the latter charged into a bed. The variety of activated carbon affects what is adsorbed and how well. Activated carbon with a concentration of small pores tends to adsorb smaller molecules than large-pored carbons. The chemistry of the carbon surface and its ash constituents also affect behavior.

The major application division of the carbon adsorption unit operation is between liquid-phase adsorption and gaseous-phase adsorption. Gaseous-phase carbon adsorption is primarily used for solvent vapor recovery and selective gas separations. Liquid-phase carbon adsorption is used to decolorize or purify liquid, solutions, and liquefiable materials such as waxes, and in water and wastewater treatment as a polishing removal or tertiary treatment.

LIQUID-PHASE ADSORPTION

Carbon adsorption from the liquid phase is generally classified as a nonpolar or hydrophobic type of adsorption operation. It is generally used to remove less polar contaminants from polar bulk streams. The two basic liquid-phase equipment designs for carbon adsorption are the fixed-bed and pulsed-bed arrangements. The equipment makes use of either powdered or granular liquid-phase carbons. Fixed-bed equipment can assume the form of either single or multiple columns which can operate in series, parallel, or both.

Contact-Batch Operation

In a typical batch configuration for a contact-batch operation, the equipment consists of an agitated tank constructed of materials suitable for the liquids being processed. The agitation allows the carbon particles to continually contact fresh portions of liquid causing mild turbulence. Where materials are sensitive to oxidation which can be caused by excessive mixing, the adsorption should be conducted under a partial vacuum, or an inert atmosphere. The liquid-carbon mixture is pumped from the tank through a filter (commonly a plate and frame type). Powdered carbon is normally applied as a slurry to minimize dusting problems.

Fixed Single-Column System

The single-column system for liquid-phase carbon adsorption is used in situations where the following conditions prevail: Laboratory testing has indicated that the breakthrough curve will be steep; the extended lifetime of the carbon at normal operating conditions results in minor replacement or regeneration costs; the capital cost of a second or third column cannot be justified due to insufficient savings in carbon cost; to preserve product qualities; or unusual temperatures, pressure, and so on must be maintained in the column. Unless any of these conditions prevail, a multiple-column adsorber is preferable because of the operating flexibility it provides.

Multiple-Column System

The choice of a multiple-column system is applicable when the nature of the process does not allow for interruption during loading, unloading, or regeneration, especially when an alternate unit is not available. Multiple-column systems are also preferable when space constraints do not allow for a single column of adequate height or residence time.

The columns of a parallel-column system are onstream at even time intervals and the column discharge is to a common manifold. The parallel design allows for smaller pumps, lower power requirements, and less stringent pressure specifications for columns and piping. Normally the carbon is not completely spent at the point where it is removed for regeneration.

The effluent from one column of a series-column system becomes the feed for the next column. The series layout is preferred over the parallel layout if the highest possible effluent purity is desired and the breakthrough curve is gradual, or the combination of a gradual breakthrough curve and high carbon demand per unit of production economically needs to exhaust the carbon. The carbon in the lead column is removed during regeneration and new carbon is put in the column onstream at the end of the series. The former lead column is replaced by the second column. The result is that normally the operating costs for series-column systems are lower than for single-column or parallel-column systems in the same application. Where the two layouts are combined in a combined series-and-parallel system, the best characteristics of each layout are realized.

The fixed-bed systems described can either have upward or downward liquid flows. Downflow operation has more of an inherent filtering capability. Suspended solids will be removed by the finer carbon particles at the top of the bed. The capture of significant quantities of suspended solids can lead to high-pressure drops. At this point the procedure is to backwash the adsorber which can take time and use significant wash liquid. Therefore, the downflow operation must have piping in both directions. The direction of flow is the same during the adsorption and washing cycles for the upflow operations. The washing cycles are far less frequent, pressures drops are lower, and considerably less downtime and wash liquid are consumed. Although some filtration will occur, upflow operations will not produce an effluent free of turbidity or suspended solids.

Pulsed-Bed Adsorbers

The carbon moves countercurrent to the liquid in pulsed-bed adsorbers. The effect is of a number of stacked, fixed-bed columns operating in series. Spent carbon is removed from the bottom of the columns as the liquid flows upward and fresh or segmented carbon is supplemented at the top. Pulsed-bed columns are usually operated with the columns completely filled with carbon, which does not allow for bed expansion during operation or cleaning. Where the pulsed-bed unit does permit bed expansion, the efficiency of the unit deteriorates due to the mixing carbon disturbing the adsorption zone. The withdrawn carbon may contain spent and partially spent carbon.

Pulsed-bed adsorbers are most commonly operated on a semicontinuous basis. During this type of operation, a set quantity of spent carbon is removed at defined intervals from the bottom of the column. Replacement of carbon is at the top. Pulsed-bed systems are the liquid-phase carbon adsorption systems that come closest to completely exhausting the carbon with the least capital investment.

The goal of either system (pulsed or fixed bed) is to maximize the use of the carbon by regenerating just the carbon that is expended. The choice of a pulsed-bed system is generally made when the feed does not contain suspended solids and the usage rate for carbon is high. Pulsed-bed systems are not effective for biologically active feeds. Conversely, fixed-bed systems are normally employed when the liquid contains significant quantities of suspended solids, is biologically active, or carbon usage is low.

Regeneration

Various alternative regeneration techniques for restoring spent carbon to its original adsorptive capacity make use of thermal, biological, chemical, hot-gas, or solvent techniques. Multihearth or rotary furnaces can be used to volatilize and carbonize adsorbed materials. Aerobic, anaerobic, or both types of bacteria can be used on site to remove adsorbed biodegradable material. Some methods are destructive to the adsorbate and no recovery can be made. Chemical, hot-gas, steam, or solvent regeneration are nondestructive methods for recovery of materials. All are carried out in place and rely on the varying adsorptive capacity of carbon for organics under changing process conditions such as pH, temperature, and nature of the liquid phase. Chemical regeneration uses a regenerant such as formaldehyde to react with the sorbed material and remove it from the carbon. Hot-gas regeneration is used when carbon has adsorbed a low-boiling point organic material. Steam, CO_2, or N_2 are passed through the bed causing vaporization. Solvent regeneration employs a suitable solvent to pass through the spent carbon and dissolve the adsorbed material. Solvents are then recycled and purified, usually through decantation or batch distillation. Chemical and solvent regeneration methods can be combined effectively. Steam is widely used for low-temperature regeneration.

Labor Requirements

For liquid-phase carbon adsorption, labor requirements range from the more labor-intensive role of a batch-sequence operator to the more continuous liquid-phase adsorption. Batch-sequence operators typically perform the following sequence of operations. The adsorption vessel is charged with the liquid process stream. Activated carbon is added in a variety of possible ways. If charging is automated or dust controlled, the operator runs this equipment. The operator typically runs the agitator for a set time cycle, then activates a pump to draw the treated mixture through a filter to remove the carbon. The attention required during the filtration cycle is highly dependent on the size of the operation and the filter capacity. After filtering, the operator washes and dries the cake as needed and removes it from the vessel. For small batch operations, the filter-cake removal operation is manual in most cases. The operator then packages the carbon in an acceptable manner for disposal in nonregenerable operations. In a regenerable operation the operator may have the responsibility of charging the spent carbon to the regeneration operation, such as a multihearth or rotary furnace in the case of heat regeneration.

For a continuous liquid-phase adsorption system, less direct operator presence than for batch adsorption is required. More of the sequences are automatically controlled and the continuous nature of the operation allows a supervisory role for the operator. Where the adsorption chambers are periodically emptied and recharged with fresh carbon on a regular basis, then additional labor is needed. The actual labor requirement of adsorption operations is a function of the total system design which includes all associated processing facilities. For both batch and continuous operations it is, to an extent, determined by the magnitude of the operation (number of units, size of equipment, complexity of operation) and the nature of the total system design which could, for example, range from a small pharmaceutical production to a large sugar refinery operation. On a case-by-case basis, certain decisions must be made, as to whether the adsorption system alone is being considered in analyzing labor needs or whether associated procedures (that is, filtration) should be considered.

Industrial Applications

The following are typical industrial applications for liquid-phase carbon adsorption. Generally liquid-phase carbon adsorbents are used to decolorize or purify liquids, solutions, and liquefiable materials such as waxes. Specific industrial applications include the decolorization of sugar syrups; the removal of sulfurous, phenolic, and hydrocarbon contaminants from wastewater; the purification of various aqueous solutions of acids, alkalies, amines, glycols, salts, gelatin, vinegar, fruit juices, pectin, glycerol and alcoholic spirits; dechlorination; the removal of grease from dry-cleaning solvents and from electroplating solutions; and the removal of wastes, aniline, benzene, phenol, and camphor from water. Adsorption for the removal of trace contaminants is widely used commercially for the recovery of major components of feed streams as pure products.

GASEOUS-PHASE ADSORPTION

Gaseous-phase carbon adsorption systems can be classified in several ways. The first category is between regenerable and nonregenerable processes. The majority of industrial systems are regenerable operations that allow the user to recover the adsorbate and continue to reuse the activated carbon adsorbent.

Regeneration relies on the continuity of gaseous adsorption achieved through equipment cycling to a desorption or regeneration phase of operation in which the temporarily exhausted beds of carbon are regenerated by removing the adsorbate. Regeneration operations are categorized in the following mechanisms: thermal-swing regeneration, pressure-swing regeneration, inert-gas purge stripping, and displacement desorption.

Thermal swing is widely used regeneration in purification adsorption operations. The spent bed is heated to a level at which the adsorptive capacity is reduced so that the adsorbate leaves the activated carbon surface and is removed in a stream of purge gas. *Pressure swing* relies on the reduction of pressure at constant temperature to reduce the adsorptive capacity for an adsorbate. Pressures can drop from elevated to atmospheric or from atmospheric to vacuum conditions. *Inert purge stripping* relies on the passage of a liquid or gas, without adsorbable molecules and in which the adsorbate is soluble, through the spent carbon bed at constant temperature and pressure. *Displacement desorption* relies on the passage of a fluid containing a high concentration of an adsorbable molecule or a more strongly adsorbable molecule than the adsorbate presently on the carbon.

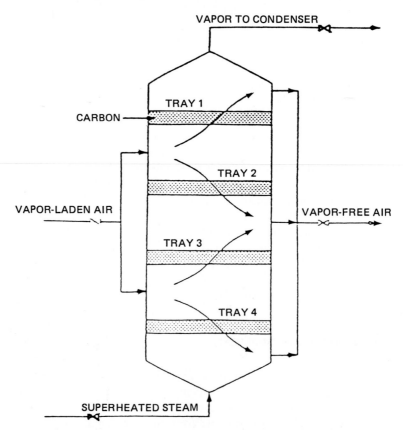

Figure 8.1 Cross section of adsorber with four fixed beds of activated carbon.

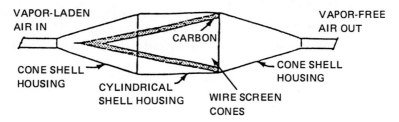

Figure 8.2 Horizontal adsorber.

Gaseous-phase adsorption systems are also categorized as either fixed-bed adsorbers or movable-bed adsorbers.

Fixed-Bed Adsorbers

The various configurations of fixed-bed gaseous-phase carbon adsorption systems are illustrated in various sections throughout this book as well as Figures 8.1, 8.2, and 8.3. Enclosures for simple fixed-bed adsorbers may be vertical

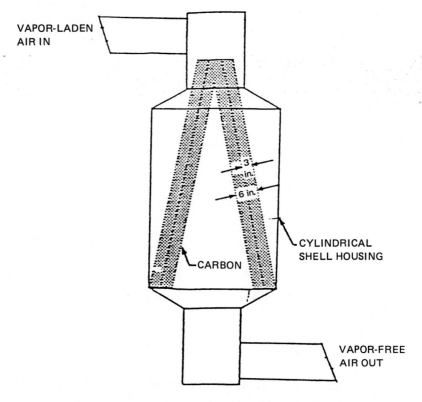

Figure 8.3. Diagrammatic sketch of vertical adsorber with two cones, permitting studies on different depths of carbon beds.

or horizontal, cylindrical, or conical shells. Where multiple fixed beds are needed, the usual configuration is a vertical cylindrical shell. The type of enclosure used is normally dependent on the gas volume handled and the permissible pressure drop.

The gas flow can be either down or up. Downflow allows for the use of higher gas velocities, while in upflow the gas velocity must be maintained below the value which prevents carbon boiling which damages the bed. When large volumes of gas need to be handled, cylindrical horizontal vessels are selected. The beds are oriented parallel to the axis.

For the continuous operation of fixed-bed adsorbers, it is desirable to have two or three units. With two adsorber units, one unit adsorbs while the other regenerates or desorbs. The required times for regeneration and cooling of the adsorbent are the factors determining the cycle time. Under most situations, two adsorbing units are sufficient if the regeneration and cooling of the second bed can be completed prior to the breakthrough of the first unit. The move to three units makes it possible for one bed to be adsorbing, one cooling, and the third regenerating. The vapor-free air from the first bed is used to cool the unit which was just regenerated. Occasionally a fourth bed is used. An example arrangement of four beds would be to have two units adsorbing in parallel, discharging exhaust to a third unit on the cooling cycle as a fourth unit is regenerated. Figure 8.4 shows a diagrammatic sketch of a two-unit fixed-bed adsorber. A three-bed adsorber configuration is illustrated in Figure 8.5.

Conical fixed-bed adsorbers are employed when a low-pressure drop is desired. For systems of the same diameter and same weight of carbon, the

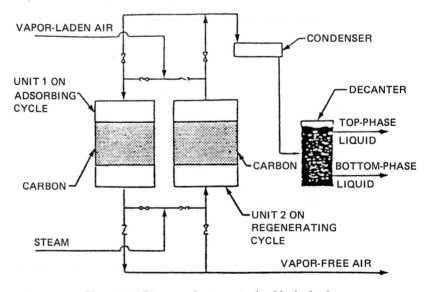

Figure 8.4 Diagram of a two-unit, fixed-bed adsorber.

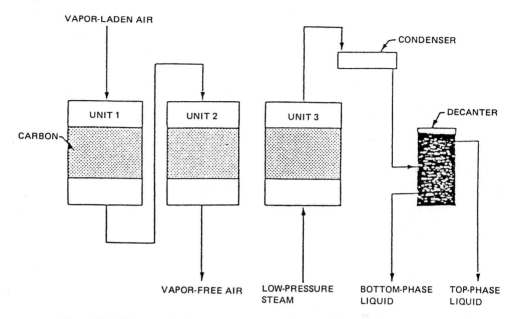

VAPOR-LADEN AIR

CONDENSER

DECANTER

CARBON

UNIT 1   UNIT 2   UNIT 3

VAPOR-FREE AIR   LOW-PRESSURE   BOTTOM-PHASE   TOP-PHASE
                  STEAM   LIQUID   LIQUID

Figure 8.5   Diagram of a three-unit operation of a fixed-bed adsorber showing No. 1 and No. 2 adsorbing in a series and No. 3 regenerating. Second cycle, No. 2 and No. 3 will be adsorbing with No. 1 regenerating. Final cycle, No. 3 and No. 1 will be adsorbing with No. 2 regenerating.

pressure drop through a cone-shaped bed is less than one half of that through a conventional flat-bed adsorber, while the air volume is more than doubled that through the flat bed.

Movable-Bed Adsorbers

These are primarily used for solvent recovery and consist of a totally enclosed rotating drum housing which encloses a bed of activated carbon. A fan delivers solvent-laden air into the enclosure through ports and into the carbon section above the bed. The solvent-laden air passes through the bed to a space on the inside of the cylindrical carbon bed. The clean air discharge is through ports at the end of the drum opposite the entrance, axially to the drum and out to the atmosphere. Steam is normally used to regenerate the movable-bed adsorbers. A continuous carbon adsorber is diagrammed in Figure 8.6.

In a typical start-up of a gas adsorption system, the operator conducts a general check of all system components: gaskets, bypass valves, adsorber and alternative unit, temperatures, and time clocks. Desired cycle times are set controlling adsorption times, purge time, hot-gas drying and cooling. Following the start of gas flow, pressure drops and temperature rises are

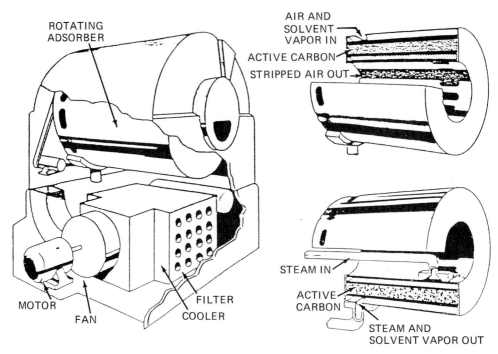

ROTATING
ADSORBER

AIR AND
SOLVENT
VAPOR IN
ACTIVE CARBON
STRIPPED AIR OUT

MOTOR
FAN
FILTER
COOLER

STEAM IN
ACTIVE
CARBON
STEAM AND
SOLVENT VAPOR OUT

Figure 8.6   *(Left)*: A rotating fixed-bed continuous adsorber showing the path of the vapor-laden air to the carbon bed. *(Right)*: A continuous adsorber showing path of steam during regeneration.

monitored. In the case of multiple units, the operator checks to see that desorption is proceeding properly, the stripping fluid is turned on, the regeneration or disposal system for the dissolved vapors is checked, the condenser cooling-water lines are opened, and the temperature of the cooling-water exit temperature is monitored.

During operation, temperature and pressure drops are monitored. The prefilter is monitored to insure adequate gas flow to the processing system and is changed when a high-pressure drop is recorded. A check is made on the bed to be regenerated to make sure the proper carbon bed sequencing is set and the bed is prepared for the steam or stripping medium. Where a third bed is used, a check is made to insure that this is being properly cooled. There should be a provision for routinely monitoring the adsorber emissions. Although a proper cycling procedure has been used between alternate beds, an unexpected contamination of the adsorbent would cause a premature breakthrough of the beds resulting in the release of contaminants. A routine shutdown would normally only involve the shut off of the gas flow from the process. If the adsorber were to be shut down for a lengthy period of time, complete draining of all lines and vessels normally would be practical.

The maintenance duties are relatively simple. Frequent inspection for equipment abrasion and corrosion is part of a preventive maintenance program. Filters need to be inspected frequently to check for holes or plugging, and when necessary they are cleaned and/or replaced. Other maintenance duties consist of maintaining adequate oil levels in lubrication reservoirs; inspecting the general condition of the exterior; periodic testing for bed contaminants; inspecting/repairing pumps blowers, valves, and so on; inspecting recovery system controls including cooling-water temperature, condenser temperatures, and so on; inspecting built-in safety devices; and inspecting the regeneration system. The frequency of the preventive maintenance varies from daily to monthly depending on the application and the manufacturer's suggestions.

Industrial applications for gaseous-phase carbon adsorption have been discussed. The major use of carbon adsorption for gaseous systems is in solvent recovery from exhaust leaving a process evaporation chamber. This type of use is employed in graphic arts operations, various types of spray painting applications, textile dry-cleaning operations, and polymer processing. Carbon adsorption is used extensively in the control of vapors from petroleum marketing plants. While regulations limit the emission of volatile organic compounds (VOCs), most of these gas adsorption operations would be practical anyway because of the value of recovered solvent. Less common applications include the fractionation of gases, low molecular weight hydrocarbons, rare gases, and industrial gases, and the purification of intake gases, circulating air, or process exhaust air to remove odors, toxic gases, and so on.

ADSORPTION THEORY

A gas or vapor when brought into contact with a solid substance has the tendency to collect on the surface of the solid. The phenomenon is known as adsorption. The amount of adsorption on the surface of most solids is exceedingly small but materials such as activated alumina, silica gel, and activated carbon have been developed to adsorb gases and vapors on their surfaces. These materials are porous solids and have an unusually high surface development in the form of an ultramicroporous structure, thus possessing a very large internal surface. A fluid is able to penetrate through the pore structure of these materials and be in contact with the large surface area available for adsorption.

The mechanism by which this surface adsorption takes place is complex. Many theories have been offered to explain adsorption, details of which may be found in the literature. The important types of adsorption are physical adsorption, in which case the gas is attracted to the surface of the adsorbent, and chemical adsorption in which the gas shows a strong interaction with the manner of a chemical reaction. The surface attraction is due to van der Waals' forces, the intermolecular forces that produce normal condensation to the liquid state. On smooth surfaces, the van der Waals' adsorption is re-

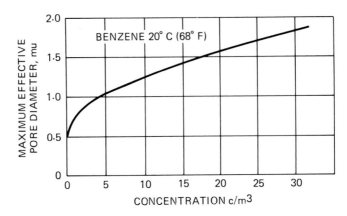

Figure 8.7 Relationship between pore size and vapor concentration.

stricted to a layer of not more than a very few molecules in thickness, but in a porous solid with a capillary structure, the surface adsorption is supplemented by capillary condensation which is also brought about by the van der Waals' forces of attraction.

Adsorptive processes are exothermic. The heat of adsorption due to surface attraction is rather greater than the heat of condensation of the gases being adsorbed. The heat of chemical adsorption increases progressively with increases in partial pressure of gas. At low partial pressure adsorption is by surface attraction, and at higher partial pressures the smallest wetted capillaries become effective and condensation begins. At the higher pressure the larger capillaries become effective.

Figure 8.7 shows the relationship between the maximum effective pore size and the vapor pressure (or vapor concentration) for benzene at 20°C. These sizes are computed on the basis of the capillary condensation theory. The relationship between pressure and amount adsorbed is dependent on the size distribution of the capillary pores as well as the area of the exposed surface and the nature of adsorbent and gas.

These relationships can be expressed graphically in the form of adsorption isotherms, determined experimentally, and are expressions of the amount of gas adsorbed under true static equilibrium conditions.

ADSORPTION ISOTHERMS

Various types of isotherms are observed and the shapes of the graphs vary according to the adsorption system. According to Brunauer, five types of adsorption isotherms exist in the literature on the adsorption of gases, and these are shown in Figure 8.8. Adsorption is specific depending on the nature of the system. Preferential adsorption is of importance for the selective re-

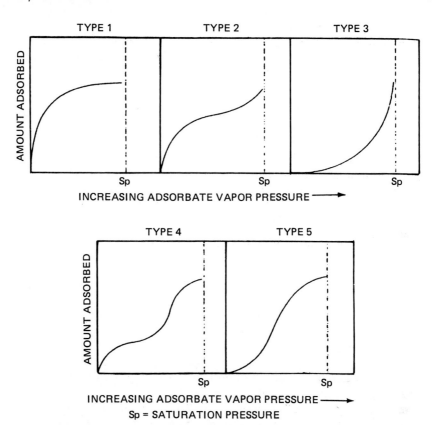

Figure 8.8 Types of adsorption isotherms: type 1, adsorption of oxygen on carbon at $-183°C$ ($-298°F$); type 2, nitrogen on iron catalyst at $-195°C$ ($-319°F$); type 3, bromine on silica gel; type 4, benzene on ferric oxide gel at 50°C (122°F); type 5, water vapor on carbon 100°C (212°F).

moval of compounds from fluid mixtures. When mixtures are adsorbed, the presence of each affects the equilibrium of the others. In general, a molecule of high molecular weight, high critical temperatures, and low volatility is adsorbed in preference to a molecule of low molecular weight, low critical temperature, and high volatility. A preferentially adsorbed molecule will displace others which have already been adsorbed.

Use of this preferential adsorptive property is made in the selective adsorption, by actuated carbon of a single hydrocarbon gas from a mixture. In such applications the activated carbon may be regarded as a rectification plant. By adjustment of the adsorption period, any particular cut can be produced from the fractions contained in the gas. For example, in a hydrocarbon mixture, methane which is adsorbed first will be displaced by ethane, which in turn will be displaced by the propane, and this process of displacement will continue throughout the series of hydrocarbons.

DESORPTION OR REGENERATION

Regeneration, desorption, or stripping of the adsorbed gases from an adsorbent may be accomplished in a number of ways:

- The temperature of the adsorbent may be raised until the vapor pressure of the adsorbed gas exceeds atmospheric pressure. The adsorbed gas will then be evolved and may be collected at atmospheric pressure.
- Adsorbed gas may be withdrawn by vacuum application lowering the pressure below the vapor pressure of the adsorbed gas.
- Adsorbed gas may be withdrawn in a stream of inert gas passing through the adsorbent, keeping the partial pressure of the stripped gas in the gas stream below the equilibrium pressure of the adsorbed gas.
- Adsorbed gas may be withdrawn in a stream of an easily condensable gas such as steam. Stripped gas is recovered by condensing the stripped gas and steam mixture.
- Adsorbed gas may be displaced by the adsorption of a gas which is preferentially adsorbed.

Manufacture of Activated Carbon

Activated carbon may be manufactured from a wide range of carbonaceous substances including bones, coals, wood dust, peat, nutshells, and wood charcoal. The fine capillary structure is formed during the activation process. The raw material is a compound of amorphous carbon and hydrocarbons not active but becoming active when the hydrocarbons held in the carbon are removed by oxidation. The combined effect produces an ultrafine capillary structure throughout the carbon. The principal commercial methods used are chemical activation and steam activation.

The chemical method of activation consists of mixing a pulverized form of carbonaceous material with a liquid dehydrating agent, drying the mixture, and heating it in a retort to complete the activation. The most widely used chemicals are zinc chloride and phosphoric acid.

Raw material for steam activation is usually a carbonized material derived from wood, peat, brown coal, and so on. The charcoal is heated in a retort to a high temperature. Steam, the activating agent, is passed through the bed of heated charcoal to produce the desired porous structure, and the resulting water gas produced is often used for heating the retorts.

Quality and characteristics of activated carbon depend on the physical properties of the raw materials and methods of activation used. There is a wide choice of raw materials and control of the activation process to produce activated carbons having widely varying physical and adsorptive properties.

Differing physical properties of the many types and grades of activated carbon commercially available make selection difficult. It is therefore advisable to discuss the application with the carbon manufacturer in order to obtain the carbon grade best suited for the intended application. Most manufacturers are willing to cooperate and offer useful information regarding the application of their materials.

Properties of Activated Carbon

The great advantage activated carbon has over other recovery systems is the outstanding ability to recover organics from low concentrations easily and inexpensively. This is desirable in processes using inflammable or toxic compounds, where it is necessary for safety reasons to ensure adequate ventilation to prevent solvent concentrations from reaching dangerous proportions.

Gas purification applications discussed in Chapters 2 and 4 involving the removal of small concentrations of impurities include the deodorization of air, removal of traces of organic impurities from gas streams to prevent catalyst poisoning, removal of traces of oil vapors from compressed gases, and the removal of similar substances from fluid streams. Important gas separations using activated carbon are discussed in subsequent chapters.

OTHER ADSORBENTS

Silica Gel

Silica gel is a granular adsorbent having a translucent appearance. Manufacture consists essentially of adding a solution of sodium silicate to sulfuric acid, washing the gel with alcohol, drying, roasting, and grading. Silica gel shows a specific selective adsorption for water vapor which is higher than either activated carbon or activated alumina, and hence its principal application is for the dehydration of gases. It is capable of adsorbing up to 40 percent of its weight of water vapor and may be simply reactivated by passing heated gas or air through the adsorbent and cooling.

Activated Alumina

Aluminum oxide-base adsorbents are prepared by heat treatment of bauxite or alumina hydrate, producing a porous solid adsorbent. Activated alumina has the higher adsorbent capacity but because of its lower cost, activated bauxite has found desiccant applications in competition with silica gel and activated alumina. Both activated alumina and activated bauxite are also widely used for the dehydration of gases. Reactivation is carried out by passing heated gas or air through the spent adsorbent and cooling.

Preferential adsorption characteristics and physical properties of the industrial adsorbents determine the main applications for each type. All adsorbents are capable of adsorbing organic solvents, impurities, and water vapor from gas streams, but each has a particular affinity for water vapor or organic vapors.

Activated alumina, silica gel, and molecular sieves will preferentially adsorb water from a gas mixture containing water vapor and organic solvent. This is a serious disadvantage in solvent recovery work where the water content of the air or gas stream is often greater than the solvent content.

Silica gel and activated alumina disintegrate under liquid water, rendering their use for the desorption of organic solvents much more difficult. Such adsorbents, therefore, are normally used only for the drying of air and gases and are regenerated by blowing a stream of hot air or gas through the adsorbent bed. Carbon is normally regenerated by direct steam to facilitate the collection of the stripped solvents by simply condensing the steam-solvent vapors produced.

CARBON ADSORPTION APPLICATIONS

Carbon has been known throughout history as an adsorbent with its usage dating back centuries before Christ. Ancient Hindus filtered their water with charcoal. In the thirteenth century, carbon materials were used in a process to purify sugar solutions. In the eighteenth century, Scheel discovered the gas adsorptive capabilities of carbon and Lowitz noted its ability to remove colors from liquids. Carbon adsorbents have been subjected to much research resulting in numerous development techniques and applications.

One of these applications was begun in England in the mid-nineteenth century with the treatment of drinking waters for the removal of odors and tastes. From these beginnings, water and wastewater treatment with carbon has become widespread in municipal and industrial processes, including wineries and breweries, paper and pulp, pharmaceutical, food, petroleum and petrochemical, and other establishments of water usage. Interest in carbon use for air as well as water pollution control and traditional industrial/product applications has received increased attention since the early 1970s with the advent of more stringent environmental regulations.

ADSORPTION PROCESS

The adsorption process occurs at solid-solid, gas-solid, gas-liquid, liquid-liquid, or liquid-solid interfaces. Adsorption with a solid such as carbon depends on the surface area of the solid. Thus, carbon treatment of water involves the liquid-solid interface. The liquid-solid adsorption is similar to the other adsorption mechanisms. There are two methods of adsorption: physisorption and chemisorption. Both methods take place when the molecules in the liquid phase become attached to the surface of the solid as a result of the attractive forces at the solid surface (adsorbent) overcoming the kinetic energy of the liquid contaminant (adsorbate) molecules.

Physisorption occurs when, as a result of energy differences and/or electrical attractive forces (weak van der Waals' forces), the adsorbate molecules become physically fastened to the adsorbent molecules. This type of adsorption is multilayered; that is, each molecular layer forms on top of the previous layer with the number of layers being proportional to the contaminant concentration. More molecular layers form with higher concentrations of contaminant in solution.

When a chemical compound is produced by the reaction between the adsorbed molecule and the adsorbent, chemisorption occurs. Unlike physisorption, this process is one molecule thick and irreversible because energy is required to form the new chemical compound at the surface of the adsorbent, and energy would be necessary to reverse the process. The reversibility of physisorption is dependent on the strength of attractive forces between adsorbate and adsorbent. If these forces are weak, desorption is readily effected.

Factors affecting adsorption include:

- The physical and chemical characteristics of the adsorbent, that is, surface area, pore size, chemical composition, and so on.
- The physical and chemical characteristics of the adsorbate, that is, molecular size, molecular polarity, chemical composition, and so on, and the concentration of the adsorbate in the liquid phase (solution).
- The characteristics of the liquid phase, that is, pH, temperature.
- The residence time of the system.

ADSORPTION WITH ACTIVATED CARBON

Certain organic compounds in wastewaters are resistant to biological degradation and many others are toxic or nuisances (odor, taste, color forming), even at low concentrations. Low concentrations may not be readily removed by conventional treatment methods. Activated carbon has an affinity for organics and its use for organic contaminant removal from gaseous streams and wastewaters is widespread.

The effectiveness of activated carbon for the removal of organic compounds from fluids by adsorption is enhanced by its large surface area, a critical factor in the adsorption process. The surface area of activated carbon typically can range from 450–1,800 m^2/g. Some carbons have been known to have a surface area up to 2,500 m^2/g, and examples are shown in Table 8.1.

Of less significance than the surface area is the chemical nature of the carbon's surface. This chemical nature or polarity varies with the carbon type and can influence attractive forces between molecules. Alkaline surfaces are characteristic of carbons of vegetable origins and this type of surface polarity affects adsorption of dyes, colors, and unsaturated organic compounds. Silica

TABLE 8.1 SURFACE AREAS OF SOME TYPICALLY AVAILABLE ACTIVATED CARBONS

Origin	Surface Area (m^2/g)
Bituminous coal	1,200–1,400
Bituminous coal	800–1,000
Coconut shell	1,100–1,150
Pulp mill residue	550–650
Pulp mill residue	1,050–1,100
Wood	700–1,400

gel, an adsorptive medium that is not a carbon compound, has a polar surface which also exhibits an adsorptive preference for unsaturated organic compounds as opposed to saturated compounds. However, for the most part, activated carbon surfaces are nonpolar, making the adsorption of inorganic electrolytes difficult and the adsorption of organics easily effected.

ACTIVATED CARBON CHARACTERISTICS

Pores of the activated carbon exist throughout the particle in a manner illustrated in Figure 8.9. The pore structure of activated carbon affects the large surface-to-size ratio. The macropores do not add appreciably to the surface area of the carbon, but provide a passageway to the particle interior and the micropores. The micropores are developed primarily during carbon activation and result in the large surface areas for adsorption to occur.

Macropores are those pores greater than 1,000 A; micropores range between 10–1,000 A. Pore structure, like surface area, is a major factor in the adsorption process. Pore-size distribution determines the size distribution of molecules that can enter the carbon particle to be adsorbed. Figure 8.10 illustrates the discriminatory practices of the pores. Large molecules can block

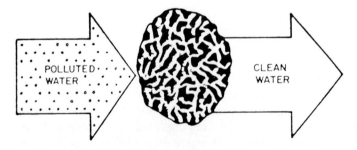

Figure 8.9 Artist's conception of a carbon granule. Organics along with the water pass through the pores and become adsorbed on the pore surfaces.

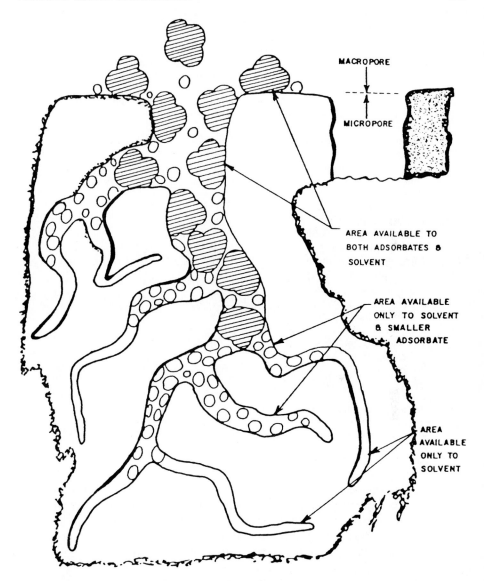

Figure 8.10 Artist's conception of molecular discrimination effects of carbon pores.

off micropores rendering useless their available surface areas. However, because of irregular shapes and constant molecule movement, the smaller molecules usually can penetrate to the smaller capillaries.

Since adsorption is possible only in those pores that can be entered by molecules, the carbon adsorption process is dependent on the physical characteristics of the activated carbon and the molecular size of the adsorbate. Each application for carbon treatment must be cognizant of the characteristics

TABLE 8.2   PROPERTIES OF SEVERAL COMMERCIALLY AVAILABLE GRANULATED CARBONS

	ICI America Hydrodarco 3000	Calgon Fitrasorb 300 (8 × 30)	Westvaco Nuchar W-L (8 × 30)	Witco 517 (12 × 30)
Physical Properties				
Surface Area, m²/gm (BET)	600–650	950–1,050	1,000	1,050
Apparent Density, g/cc	0.43	0.48	0.48	0.48
Density, Backwashed and Drained, lb/ft³	22	26	26	30
Real Density, g/cc	2.0	2.1	2.1	2.1
Particle Density, g/cc	1.4–1.5	1.3–1.4	1.4	0.92
Effective Size, mm	0.8–0.9	0.8–0.9	0.85–1.05	0.89
Uniformity Coefficient	1.7	≤1.9	≤1.8	1.44
Pore Volume, cc/g	0.95	0.85	0.85	0.60
Mean Particle Diameter, mm	1.6	1.5–1.7	1.5–1.7	1.2
Specifications				
Sieve Size (U.S. standard series)				
Larger than No. 8, max. %	8	8	8	[b]
Larger than No. 12, max. %	[b]	[b]	[b]	5
Smaller than No. 30, max. %	5	5	5	5
Smaller than No. 40, max. %	[b]	[b]	[b]	[b]
Iodine No.	650	900	950	1,000
Abrasion No., minimum	[c]	70	70	85
Ash, %	[c]	8	7.5	0.5
Moisture as packed, max. %	[c]	2	2	1

[a] Other sizes of carbon are available on request from the manufacturers.
[b] Not applicable to this size carbon.
[c] No available data from the manufacturer.

of the contaminant to be removed and designed with the proper carbon type in order to attain optimum results. Table 8.2 gives the properties of some commercially available granulated activated carbons.

Basically, there are two forms of activated carbon: powdered and granular. The former are particles that are less than U.S. Sieve Series No. 50, while the latter are larger.

The adsorption rate is influenced by carbon particle size, but not the adsorptive capacity which is related to the total surface area. By reducing the particle size, the surface area of a given weight is not affected. Particle size contributes mainly to a system's hydraulics, filterability, and handling characteristics.

CARBON ACTIVATION

Carbon materials are activated through a series of processes which includes:

• Removal of all water (dehydration).

- Conversion of the organic matter to elemental carbon, driving off the noncarbon portion (carbonization).
- Burning off tars and pore enlargement (activation).

Initially, the material to be converted is heated to 170°C to effect water removal. Temperatures are then raised above 170°C driving off CO^2, CO, and acetic acid vapors. At temperatures of about 275°C, the decomposition of the material results and tar, methanol, and other by-products are formed. Nearly 80 percent elemental carbon is then effected by prolonged exposure to temperatures ranging from 400–600°C.

Activation of this product follows with the use of steam or carbon dioxide as an activating agent. The superheated steam at 750–950°C passes through the carbon, burning out by-product blockages, expanding and extending the pore network.

CARBON SYSTEMS

Water, Wastewater, and Activated Carbon

Activated carbon is commonly used in water and wastewater treatment, removing organics that cause odors, tastes, and other detrimental effects. In addition, as a recycling medium, activated carbon can be used for solvent purification or recovery of expensive materials. The economics of carbon systems have been improving such that their usage is becoming more accepted.

The utilization of carbon at a wastewater or water treatment facility can be in the form of a powder or granule. Granular carbon is placed in a bed and raw water or wastewater is passed over it. Tastes, colors, and odors are removed from potable waters, and dissolved organics such as phenols, pesticides, organic dyes, surfactants, and so on are removed from industrial and

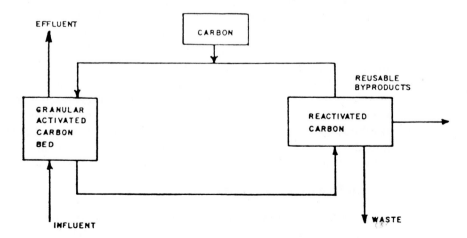

Figure 8.11 Simple schematic of a granular activated carbon-bed system.

TABLE 8.3   SEVERAL ACTIVATED CARBON REMOVAL APPLICATIONS

Acetaldehyde	Gasoline
Acetic Acid	Glycol
Acetone	Herbicides
Activated Sludge Effluent	Hydrogen Sulfide
Air Purification Scrubbing Solutions	Hypochlorous Acid
Alcohol	Insecticides
Amines	Iodine
Ammonia	Isopropyl Acetate and Alcohol
Amyl Acetate and Alcohol	Ketones
Antifreeze	Lactic Acid
Benzine	Mercaptans
Biochemical Agents	Methyl Acetate and Alcohol
Bleach Solutions	Methyl-Ethyl-Ketone
Butyl Acetate and Alcohol	Naptha
Calcium Hypochlorite	Nitrobenzenes
Can and Drum Washing	Nitrotoluene
Chemical Tank Wash Water	Odors
Chloral	Organic Compounds
Chloramine	Phenol
Chlorobenzene	Potassium Permanganate
Chlorine	Sodium Hypochlorite
Chlorophenol	Solvents
Chlorophyl	Sulfonated Oils
Cresol	Tastes (Organic)
Dairy Process Wash Water	Toluene
Delayed Organic Water	Trichlorethylene
Defoliants	Trickling Filter Effluent
Detergents	Turpentine
Dissolved Oil	Vinegar
Dyes	Well Water
Ethyl Acetate and Alcohol	Xylene

municipal wastewaters. Table 8.3 lists applications for activated carbon's adsorption abilities for various compounds.

The removal process continues until the carbon reaches its adsorption saturation point, at which time it is regenerated. The waste and recoverable products are extracted with a regeneration solution, the latter being reused and the waste discharged. Figure 8.11 is a simple schematic of a granular activated carbon-bed system.

Carbon Treatment Techniques

Powdered activated carbon is usually added to the water or wastewater stream with automatic chemical feeders. These feeders meter in the carbon at a predetermined rate at various points in the system. The point of carbon application varies with the treatment process and the desired results.

Carbon added during the early stages of treatment effects a more stable sludge and a better floc formation. In addition, because the carbon has adsorbed much of the organic matter, less chlorine disinfectant is required. Should carbon costs be high, it would be advantageous to administer the carbon later in the process stream. After sedimentation, less carbon would be required because an appreciable amount of the organic matter is removed in the flocculation process. Tertiary treatment involves the addition of carbon after filtration as a final polishing process. Best efficiencies with carbon are usually obtained with several points of application. (See Figure 8.11.)

Granulated activated carbon beds are used in a similar fashion. Like powdered carbon, carbon-bed unit processes can be implemented at various points in a treatment plant. A carbon bed, situated just after the primary treatment and chemical addition process, would not only be used to remove dissolved organics, but also biodegradable organics, and suspended solids and colloidal materials. This is known as the physical-chemical treatment (PCT) process and does not involve biological treatment, that is, activated sludge, trickling filters, and biodisks.

The physical-chemical treatment process results in cost savings since the biological treatment unit is omitted. However, increased carbon loadings from PCT result in high regeneration rates (costs), thus minimizing the capital investment savings. Further, depending on the raw wastewater loadings, PCT may not be able to effect the required effluent characteristics. Figure 8.12 illustrates various PCT flow schematics.

Carbon beds have been used effectively in conjunction with biological treatment to obtain a high-quality effluent. In following unit operations of biodegradation and filtration, carbon-bed adsorption has several advantages, including: (1) organics, BOD and COD levels are reduced and require shorter carbon-bed retention times; (2) operating costs are reduced because regeneration of the carbon is minimized and suspended solids would not clog the bed; and (3) biological growth on the carbon and its associated problems would be kept to a minimum. Figure 8.13 shows some points of application of carbon-bed unit processes after biological treatment.

The uses of activated carbon are many. Some of the advantages of an activated carbon system include:

- Considering the expense of industrial space, carbon beds require a relatively small amount of space.
- The process generates no secondary sludge.
- The process produces no odors. Recovery of valuable materials can be effected.

Wastewater streams, volume and composition fluctuate; therefore, depending on the flow requirements and wastewater characteristics, an acti-

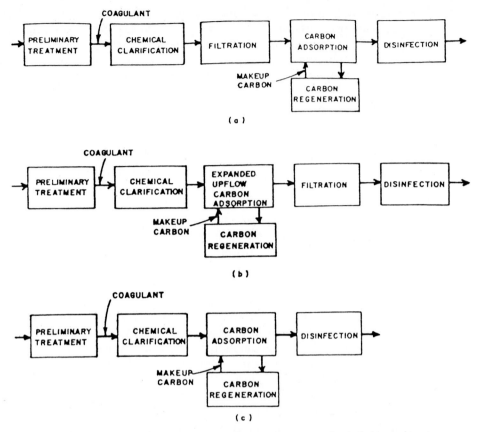

Figure 8.12 Various carbon application techniques in a physical-chemical treatment process.

vated carbon system can be designed. Should the flow of wastewater at a facility vary drastically, a system such as the adsorption system shown in Figure 8.12 could be employed. Since contact times and flow rates are dependent (adsorber size, diameter, and length), a longer residence time can be obtained by reducing the flow rate through the system. High flow rates and/or long residence times can be accommodated with the addition of several module systems in series or parallel. This type of system is best suited for use at PCT-type facilities.

There are many types of carbon systems, each with its own advantages. Depending on flow rate, flow-rate variations, wastewater characteristics, effluent requirements, application, treatment process, and economics, a particular system can be selected. With each carbon process there are the inevitable trade-offs.

A countercurrent or upflow carbon bed effects a highly efficient use of carbon, reducing regeneration and carbon makeup costs. Further, it is a con-

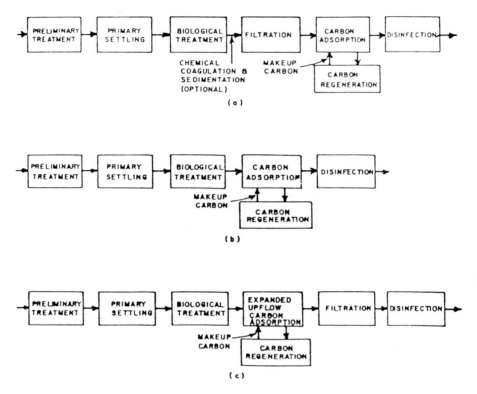

Figure 8.13 Points of application of carbon in a biological treatment plant.

tinuous operation that can be designed to eliminate downtimes for regeneration. However, this single-purpose use of carbon results in the need for other treatment units.

Concurrent or downflow carbon beds not only can remove organics by adsorption but also suspended solids by filtration. This results in a smaller capital investment because a filtration unit is no longer required. However, operating costs are increased as the carbon must be regenerated more frequently, and efficiencies are lower than if the carbon were used for a single purpose.

Another variable in a carbon-bed system would be the manner in which the wastewater is introduced to the carbon. The influent may flow through a carbon contactor due to gravity, in which case capital costs are kept to a minimum because conventional construction materials and techniques can be used. Pumps and their associated operating costs would not be required. On the other hand, with a pressurized flow through a carbon bed, a system has the capability of handling upset conditions more easily as flow and wastewater chemical composition variations can be accommodated. Figure 8.14 is a schematic of a pressurized downflow carbon bed.

A carbon bed's physical dimensions and carbon size as well as backwash

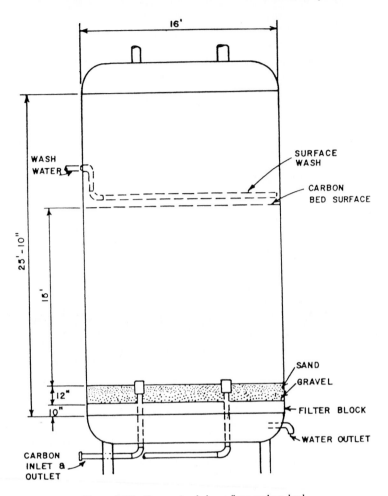

Figure 8.14 Pressurized downflow carbon bed.

requirements are subject to the containment loading of the wastewater and the flow rate. A high organically loading influent and flow rate would need a deep bed to effect a quality effluent. Generally, bed depths range from 10–30 ft and flow rates from 2–10 gpm/ft². A waste stream high in suspended-solids content would have to be backwashed more frequently than one with a low suspended-solids content. The suspended solids are filtered out of the waste stream by the carbon bed, which then begins to clog with the deposits.

Backwashing forces water through the bed in the opposite direction from normal, thereby removing the filtered debris from the carbon surfaces. Backwashing frequency may commonly be semiweekly up to a daily operation, should loadings be high.

Typical Arrangements

Columns of activated carbon are typically arranged in series to obtain a carbon-bed depth necessary to provide the required effluent quality. Wastewater flows through both as shown in Figure 8.15. When the carbon in column No. 1 is spent, column No. 2 continues to remove contaminants to maintain water quality. Column No. 1 is taken off line and, after being flushed of spent carbon and refilled with new carbon, it is tied into the system downstream of column No. 2. The process is then repeated when column No. 2 becomes saturated with impurities. Thus, this system can be used continuously, and the carbon used most efficiently.

A typical column, as noted in Figure 8.14, is designed with flat or concave tops and bottoms. A retainer screen is placed on the bottom (filter block), followed by thin layers of gravel and sand (optional), and then the carbon. Common dimensions of a carbon vessel would involve height-to-diameter ratios of 2:1, but a higher ratio would improve a system's performance.

The release of the spent carbon from the above vessels requires either a nozzle for hydraulic discharge or a man head for manual cleaning. These methods may necessitate backwashing or flushing to remove residual carbon.

To eliminate this problem of residual carbon left in the column during carbon replacement, cone-shaped bottoms are installed on the columns. A

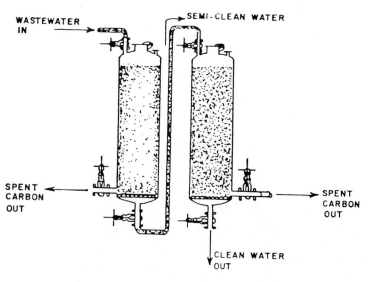

Figure 8.15 Multiple-column systems in series.

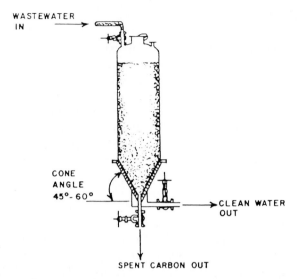

Figure 8.16   A cone-bottom column.

cone angle of 45°–60° is normally used to facilitate adequate carbon flushing (Figure 8.16).

A moving-bed system is a modified upflow carbon bed and is generally used when large amounts of carbon are required for the removal of impurities. Like a countercurrent bed, wastewater flow is up through the carbon bed

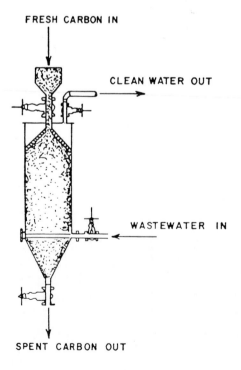

Figure 8.17   Typical moving-bed systems.

forcing the carbon particles apart and expanding in the bed. However, in a moving bed, the carbon also flows down through the column as spent carbon is periodically removed from the bottom. Fresh carbon is added at the top as the old is withdrawn to be regenerated.

Figure 8.17 illustrates this continuous-flow system which, depending on the operating parameters, can be highly efficient and economical. The carbon removed from the bottom of this system has utilized nearly all its adsorption ability; whereas in a batch operation, the carbon efficiency is low. This system is also known as a pulsed carbon bed because the carbon is not removed continuously but on an intermittent basis. Figure 8.18 illustrates typical carbon adsorption systems and their attributes.

Carbon Regeneration

The economics of carbon are such that costs would be prohibitive if it could not be reused. Thus, spent carbon passes through a regeneration or reactivation process in which organics are desorbed and the carbon can be used again. This is usually a thermal process which proceeds as follows. The exhausted granulated carbon is withdrawn from the carbon-bed column and conveyed in the form of a water slurry. Before entering a rotary kiln or multihearth furnace, the slurry is dewatered. Furnace temperatures usually range between 1,600–1,800°F during which time the carbon is dried of residual waters and the organic adsorbent is volatilized and oxidized. Combustion conditions within the furnace are controlled to limit oxygen content to effect oxidation of the adsorbed material rather than the carbon.

After thermal regeneration, the carbon is quenched in a water bath, washed of carbon fines, and recycled back to the adsorber column, or sent to storage. Air pollution devices, such as scrubbers and afterburners, are installed on the furnace to control off-gas pollutants. Figure 8.19 is a schematic diagram of the process.

During each treatment cycle, carbon losses can vary between 2 percent and 10 percent. A carbon makeup hopper and bin are included in line to provide the additional carbon necessary for the purpose. The entire thermal regeneration process usually lasts about 30 minutes.

Regeneration of carbon without heat has been effected, but usually with a recovered adsorptive capacity of low levels. However, the cost disadvantages of a low adsorptive capacity recovery level may be offset via by-product recovery.

Valuable chemicals can be recovered from a waste stream by passing them through a carbon column and then regenerating the carbon to effect product removal. An example of this is chromium. Chromium solution is introduced to a granular activated carbon column and the chromium is adsorbed by the carbon. When the carbon's adsorptive capacity is exhausted, it is regenerated. If the carbon is to be recycled, the carbon is regenerated for chromium recovery with a sodium hydroxide solution. Should the recovered

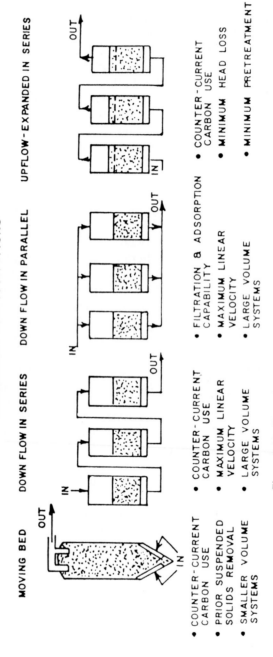

Figure 8.18 Typical carbon adsorption systems and their attributes.

GRANULAR CARBON REACTIVATION CYCLE

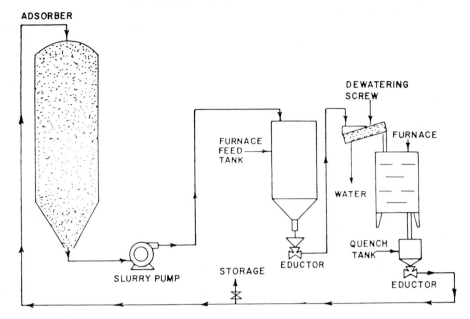

Figure 8.19 Schematic diagram of a granulated activated carbon thermal regeneration process.

adsorptive capacity of the carbon be so low that reuse would be impractical, then the chromium could be extracted from the carbon with a sulfuric acid solution. The carbon is then discarded.

The recovery of certain compounds by carbon adsorption can prove to be economically unfeasible due to high makeup carbon costs and a recovered product that may require further processing to effect purity.

Carbon Evaluation

To obtain the optimum advantage from activated carbon use, experimental analysis utilizing actual operating conditions is often necessary. The operating parameters of wastewater characteristics, treatment facility configuration, and effluent requirements are controlling factors in the selection of carbon type and mode of application.

The following list of characteristics can be evaluated for each activated carbon to determine its suitability for a particular application:

- Surface area. Generally, the larger the surface area, the more adsorption can take place.
- Apparent density. A measure of the regenerability of a carbon.

- Bulk density. Used to determine carbon quantities necessary to accomplish certain jobs.
- Effective size, mean particle diameter, uniformity coefficient. Used to determine hydraulic conditions of an adsorber column.
- Pore volume. Can be used to determine the adsorbability of a particular waste entity.
- Sieve analysis. Used to check plant-handling effects on the carbon.
- Ash percent. Shows the activated carbon's residue.
- Iodine number. An important parameter to be determined because it can indicate a carbon's ability to adsorb low molecular weights and be regenerated.
- Molasses number, value, and decolorizing index. For indication of a carbon to adsorb high molecular weights.
- Pore size. Used to obtain a carbon which can adsorb specific molecules.

The relationship between a carbon's adsorbability of a substance and that substance's concentration in a wastestream (or other liquid) is the adsorption isotherm. Adsorption isotherms are determined at constant temperatures and controlled operating conditions (pH, flow rate, and so on). From the adsorption isotherm, the amount of carbon required to effect the desired effluent characteristics can be estimated.

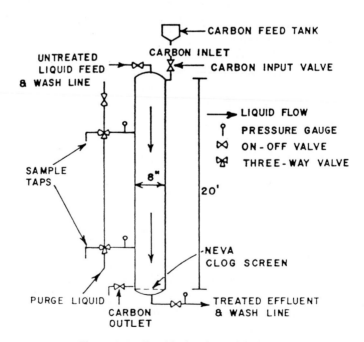

Figure 8.20 Fixed-bed carbon pilot plant.

After estimates have been determined in the laboratory, pilot plant studies should be conducted to determine:

- Carbon type, size, dosage.
- Bed dimensions.
- Effluent characteristics.
- Hydraulic characteristics.
- Dosage requirements.
- Contact time.
- Pretreatment requirements.
- Other effects.

These other effects may include bacterial growth on the carbon bed, filterability, hydrogen sulfide generation, and pH and temperature effects on adsorption. Figures 8.20 and 8.21 are examples of two configurations of granular-bed carbon pilot plants.

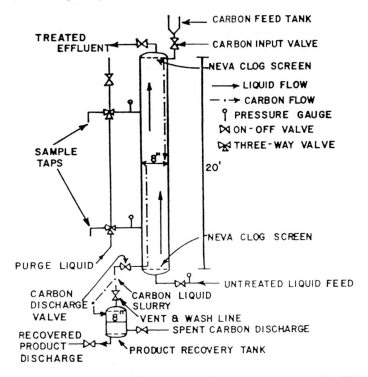

ONE FOOT SECTION OF CARBON PERIODICALLY REMOVED FROM BOTTOM WHILE AN EQUAL AMOUNT OF FRESH CARBON IS ADDED AT TOP, CARBON REMOVED IS WASHED TO RECOVER PRODUCT.

Figure 8.21 Pulsed-bed carbon pilot plant.

Breakthrough curves, a plot of wastewater constituent concentration versus wastewaters treated, are used in determining the durability of a carbon with respect to the operating conditions it will encounter at a full-scale treatment facility.

Costs

Depending on the type and mode of carbon application, capital and operating costs for the system will vary. A batch operation, such as powdered carbon added to a waste stream as illustrated in Figure 8.22, would require a significantly smaller capital investment than the construction of carbon-bed adsorbers. However, carbon applied in this manner is not recoverable and, thus, the total carbon cost over the life of the plant may prove to be uneconomical.

Carbon systems costs, like most treatment systems, are determined based on a unit cost per unit flow rate. Usually, then, as flow rates decline, unit costs increase per unit flow. The relative costs of several carbon adsorber systems are given in Table 8.4. These costs include construction, equipment, and carbon for equal treatment volumes.

As a result of an industrial spill or cleanup operation, an intermittent and intense taste and odor problem may be created. Persistent problems may be caused by the continuous discharge of process wastes. Phenols and related compounds are often the source of medicinal tastes—tastes which are intensified by clorination. Hydrocarbons from refinery wastes often can be recognized by an oily film on the water surface. As little as 0.025 to 0.050 mg

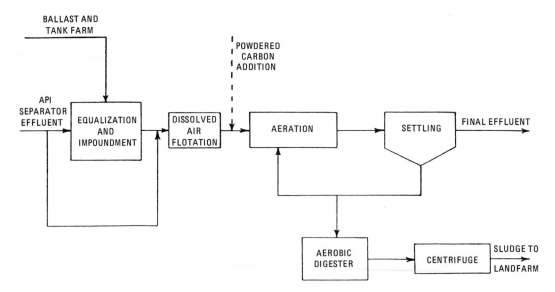

Figure 8.22 Schematic of a refinery wastewater treatment plant using powdered activated carbon.

TABLE 8.4 RELATIVE COSTS OF SEVERAL CARBON ADSORBER SYSTEMS

System	Cost Adjustment Factor
Upflow Countercurrent Packed Bed	1.0
Upflow Countercurrent Expanded Bed	1.0
Downflow-Parallel	1.0
Downflow-Series	1.28

per liter of such wastes has caused taste and odor outbreaks. Some chemical wastes, such as ethyl acrylate and N-butyl mercaptan, have been detected by the threshold odor test at 0.006 and 0.007 mg per liter, respectively. Odors resulting from pulp and paper wastes have been described as sulfite or paper-mill. Even though such terms are neither very descriptive nor precise, anyone who has been inside a paper mill will recognize the typical odor of such wastes. Wastes from the metal processing industry may cause tastes, but not odors. Zinc, copper, and other metals produce characteristic tastes.

Sewage contains a mixture of organic materials. In sewage treatment, some of these compounds may be partially oxidized and produce odor. When sewage is chlorinated to control bacteria, the effluent may have a chlorine-type odor, due to the formation of chlororganic compounds. Sewage-polluted water contains a relatively high concentration of nitrogen compounds. These may release ammonia which reacts with free chlorine to produce nitrogen trichloride. This compound has a very persistent and irritating odor.

Miscellaneous sources of taste and odor may include pest control where various chemicals have been employed, such as insecticides, fish poisons, and herbicides. Careless application has permitted some of these pesticides to enter streams which serve as sources of public water supplies. Usually the taste and odor problems have been associated with the solvents used to disperse the chemicals.

The use of activated carbon, a material specially treated to produce a surface condition of great adsorption capacity, is very effective in removing most tastes and odors. It is generally used in the form of a fine powder in dosages ranging from 10 to 50 lbs/thousand gallons. At times, as much as 300 lbs/thousand gallons are required for brief periods.

Application of carbon to the raw water will minimize the decomposition of sludge deposits in sedimentation basins. However, carbon seems to be more effective in removing taste and odor when it is on the surface of the filters where intimate contact with the filtering water is assured. Because of this, the practice has developed of adding small, uniform doses of carbon to the raw water, and adding varying and relatively larger doses to the settled water.

Many wastewater plant operators prefer to apply relatively large amounts for brief periods of time to individual filters immediately after each backwashing. This procedure has the advantage that an adequate, but not too great, amount of carbon is present on the filter bed throughout its operation period. Disadvantages are the difficulty of getting a uniform coating over the surface of the bed and the possibility of a decrease in the efficiency of taste and odor removal toward the end of the filter run.

Granular activated carbon is sometimes used as a filter, with the water passing through the beds at the rate of 2 to 4 gpm/sq ft. Since large amounts of activated carbon are quite effective in removing chlorine from water, the use of carbon filters is limited to applications before chlorination or when dechlorination as well as taste and odor removal is desired. The carbon granules become coated and the minute pores become clogged unless the water is clear. Therefore, carbon filters usually follow sand filters. Carbon filters must be washed at intervals but eventually their adsorptive power is exhausted and the carbon must be replaced or reactivated.

Adsorption with Silica Gel

Silica gel is a hard, clear, and glassy substance with a chemical composition of approximately 100 percent SiO_2, very carefully prepared in order to give a definite physical structure. It is inert chemically and strong mechanically, so that large masses of the granular material may be employed without serious loss by attrition and may be continually heated without loss to the adsorptive properties. The extremely large surface of the pores, with the addition of their capillary action and combined with its other properties, make silica gel particularly suitable for a wide variety of purposes.

Industrial applications of silica gel may be grouped under three headings:

- Adsorption from the gas phase.
- Adsorption from the liquid phase.
- As a catalyst or as a support for catalytic agents.

Silica gel selectively adsorbs vapors of volatile liquids such as water, acetone, benzene, petrol, and so on from air and similar gases. If an air-vapor mixture of this type is passed through a bed of silica gel, vapor will be adsorbed until the break point when the air leaving will contain gradually increasing quantities of the vapor. The curve shown in Figure 8.23 shows the adsorption of water vapor from air under definite conditions of temperature, humidity, and rate of flow and indicates a number of points of general interest in connection with vapor adsorption. It will be noted that under these particular conditions, the break point is reached when the silica gel has adsorbed water from the air to the extent of 20 percent of its weight. On continuing to pass air through the silica gel, water vapor will be present to an increasing extent in the air as it leaves, and the two curves shown represent alternative methods of expressing the adsorption efficiency. The average efficiency is

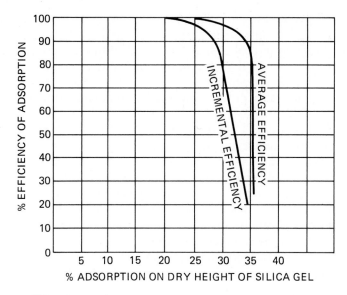

Figure 8.23 Adsorption of water vapor from air by silica gel.

calculated from the ratio of the total amount of vapor adsorbed and the total amount of vapor present in the air-vapor mixture that has passed through the gel. The incremental efficiency, on the other hand, represents the ratio of the total amount of vapor adsorbed and the total amount of vapor present in the air-vapor mixture passing through the gel at any given moment. The incremental efficiency is the actual efficiency of adsorption measured at one particular instant during an adsorption cycle, while the average efficiency is the average for the whole period of the adsorption cycle. There is an applicable difference in the adsorptive capacity between these two curves and it is important to bear this in mind when considering efficiencies. High-adsorption figures may be obtained under certain conditions, based on adsorption for complete equilibrium under static conditions, but these may be misleading and do not represent results that are of value from the industrial point of view.

A number of arrangements can be adopted employing silica gel for adsorption purposes which will depend on the specific requirements such as the quantities, concentration, temperature, and so on for any given problem. A typical plant would consist of two adsorbers, each containing the requisite quantity of silica gel, arranged so that one is adsorbing while the other is being regenerated. Regeneration will comprise removal of the adsorbed material in such a way that the silica gel has its original low condensed-vapor content and is ready for another adsorption cycle. The adsorbed liquid after removal from the gel may be condensed and recovered. Any number of adsorbers may be arranged so that they are worked in a convenient cycle of operations.

Adsorptive properties of silica gel from the gas phase afford a number of applications, such as the drying and purifying of compressed gases, dehydration of coal gas, and so on. The treatment of compressed gases, such as compressed air, oxygen, nitrous oxide, carbon dioxide, and so on, not only deals with the small residual amount of moisture remaining after compression, but also purifies the gas by removing contamination due to noxious vapors.

A special case of air drying by silica gel is the dehydration of process and instrument gases. When gas or air is passed over silica gel, the latter adsorbs water vapor and benzene. After a time, however, owing to its selective adsorptive properties, a certain amount of the adsorbed benzene may be replaced by water so that by varying the conditions it is possible that varying ratios of water and benzene adsorbed by the gel may be obtained in such application. Results obtained from a plant operation showed that when obtaining the required dehydration of some 60 percent, the ratio of water to benzene adsorbed was approximately 3:1, while at the same time, naphthalene was almost completely removed from the gas, and the very slight re-

TABLE 8.5   COMPARISON OF TREATMENT PROCESSES THAT SEPARATE ORGANICS FROM LIQUID WASTE STREAMS

Treatment Process	Required Feed Stream Properties	Characteristics of Output Stream(s)
Carbon Adsorption	Aqueous solutions; concentrations <1%; SS <50 ppm	Adsorbate on carbon; usually regenerated thermally or chemically
Resin Adsorption	Aqueous solutions; concentrations <8% SS <50 ppm; no oxidants	Adsorbate on resin; always chemically regenerated
Ultrafiltration	Solution or colloidal suspension of high molecular weight organics	One concentrated in high molecular weight organics; one containing dissolved ions
Air Stripping	Solution containing ammonia; high pH	Ammonia vapor in air
Steam Stripping	Aqueous solutions of volatile organics	Concentrated aqueous streams with volatile organics and dilute stream with residuals
Solvent Extraction	Aqueous or nonaqueous solutions; concentrations <10%	Concentrated solution of organics in extraction solvent
Distillation	Aqueous or nonaqueous solutions; high organic concentrations	Recovered solvent; still bottom liquids, sludges, and tars
Steam Distillation	Volatile organics, nonreactive with water or steam	Recovered volatiles plus condensed steam with traces of volatiles

TABLE 8.6   REPRESENTATIVES OF VARIOUS ORGANICS ON ACTIVATED CARBON

Compound	Formula	Molecular Weight	Boiling Point 760 mm C	"C" Critical Temperature	Approximate Retentivity in % at 20 C 760 mm
Methane Series	C_nH_{2n+2}				
Methane	CH_4	16.04	-184	-86.7	1
Ethane	C_2H_6	30.07	-86	32.1	1
Propane	C_3H_8	44.09	-12	95.6	5
Butane	C_4H_{10}	58.12	1	153.0	8
Pentane	C_5H_{12}	72.15	37	197.2	12
Hexane	C_6H_{14}	86.17	69	234.8	16
Heptane	C_7H_{16}	100.20	98.4	266.8	23
Octane	C_8H_{18}	114.23	125.5	296.0	25
Nonane	C_9H_{20}	128.25	150.0		25
Decane	$C_{10}H_{22}$	142.28	231.0		25
Acetylene Series	C_nH_{2n-2}				
Acetylene	C_2H_2	26.04	-88.5	36.0	2
Propyne	C_3H_4	40.06	-23.0		5
Butyne	C_4H_6	54.09	27.0		8
Pentyne	C_5H_8	68.11	56.0		12
Hexyne	C_6H_{10}	82.14	71.5		16
Ethylene Series	C_2H_{2n}				
Ethylene	C_2H_4	28.05	-103.9	9.7	3
Propylene	C_3H_6	42.08	-17.0	92.3	5
Butylene	C_4H_8	56.10	-5.0		8
Pentylene	C_5H_{10}	70.13	40.0		12
Hexylene	C_6H_{12}	84.16	64.0		
Heptlene	C_7H_{14}	98.18	94.9		25
Octalene	C_8H_{16}	112.21	123.0		25
Benzene Series	C_nH_{2n-6}				
Benzene	C_6H_6	78.11	80.1	288.5	24
Toluene	C_7H_8	92.13	110.8	320.6	29
Xylene	C_8H_{10}	106.16	144.0		34
Isoprene	C_5H_8	68.11	34.0		15
Turpentine	$C_{10}H_{16}$	136.23	180.0		32
Naphthalene	$C_{10}H_8$	128.16	217.9		30
Phenol	C_6H_5OH	94.11	182.0	419.0	30
Methyl Alcohol	CH_3OH	32.04	64.7	240.0	15
Ethyl Alcohol	C_2H_5OH	46.07	78.5	243.1	21
Propyl Alcohol	C_3H_7OH	60.09	97.19	263.7	26
Butyl Alcohol	C_4H_9OH	74.12	117.71	287.0	30
Amyl Alcohol	$C_5H_{11}OH$	88.15	138.0	307	35
Cresol	C_7H_7OH	108.13	202.5	122	30
Menthol	$C_{10}H_{19}OH$	156.26	215		20
Formaldehyde	H_3CHO	30.03	-21.9		3
Actaldehyde	CH_3CHO	44.05	21.0	188	7

duction in the properties of the dehydrated gas was more than offset by the value of the benzene recovered. Another advantage is dehydrating the gas to such an extent that water deposition in piping and tubing is prevented.

An example of the use of silica gel as a support for catalytic agents is plantinized silica gel, or gel impregnated with other suitable catalyst. This is of great value in the contact process for the manufacture of sulphuric acid and various petrochemical processes. The impregnated silica gel may be used to give rise to increased yields.

Table 8.5 gives a comparison of the capabilities of the processes to separate organics from waste streams and can be compared to carbon adsorption. Table 8.6 shows properties of various organics and retentivity on activated carbon.

9

Ion Exchange ━━

Water may contain in varying concentrations dissolved salts which dissociate to form charged particles called ions. These ions are the positively charged cations and negatively charged anions that permit the water or solution to conduct electricity and are therefore called electrolytes. Electrical conductivity is thus a measure of water purity, with low conductivity corresponding to high purity.

The process of ion exchange is uniquely suited to the removal of ionic species from water supplies for several reasons. First, ionic impurities may be present in rather low concentrations. Second, modern ion-exchange resins have high capacities and can remove unwanted ions preferentially. Third, modern ion-exchange resins are stable and readily regenerated, thereby allowing their reuse.

Other advantages ion exchange offers are:

- The process and equipment are a proven technology. Designs are well developed into preengineered units that are rugged and reliable.
- Fully manual to completely automatic units are available.
- There are many models of ion-exchange systems on the market which keep costs competitive.
- Temperature effects from 0°C to 35°C are negligible.
- The process is excellent for both small and large installations, from home water softeners to huge utility applications.

Ion exchange is a well-known method for softening or for demineralizing water. Although softening could be useful in some instances, the most likely application for ion exchange in wastewater treatment is for demineralization.

Many ion-exchange materials are subject to fouling by organics. It is possible that treatment of secondary effluent for suspended-solids removal and possibly soluble organic removal will be required before carrying out ion exchange.

Many natural materials and, more importantly, certain synthetic materials have the ability to exchange ions from an aqueous solution for ions in the material itself. Cation-exchange resins can, for example, replace cations in solution with hydrogen ions. Similarly, anion-exchange resins can either replace anions in solution with hydroxyl ions or absorb the acids produced from the cation-exchange treatment. A combination of these cation-exchange and anion-exchange treatments results in a high degree of demineralization.

Since the exchange capacity of ion-exchange materials is limited, they eventually become exhausted and must be regenerated. The cation resin is regenerated with an acid; the anion resin is regenerated with a base. Important considerations in the economics of ion exchange are the type and amounts of chemicals needed for regeneration. Often, water to be demineralized is first passed through a cation-exchange material requiring a strong acid, usually sulfuric, for regeneration. The exchange material is called a *strong acid resin*. The amount of acid regenerant is somewhat more than the stoichiometric amount, possibly 100 percent excess or more. If sulfuric acid is the regenerating acid, a waste brine is produced consisting of sulfates of the various actions in the water being treated. Because the partially treated water contains mineral acids, it is common to pass it next through an acid-absorbing resin or *weak base resin*. This resin can be regenerated with either a weak or strong base. Efficiency of regenerant use is quite high with these resins. If sodium hydroxide is the regenerating base, a waste brine is produced consisting of the sodium salts of the various anions in the water being treated. Certain anionic materials are not removed by the weak-base resin and must be further treated with strong-base resin if thorough demineralization is desired. Regenerant usage by the strong-base resins is poorer than for the weak-base resins.

IMPORTANCE OF HIGH-QUALITY WATER FOR INDUSTRIAL USE

Water problems in cooling, heating, steam generation, and manufacturing are caused in large measure from the kinds and concentrations of dissolved solids, dissolved gases, and suspended matter in the makeup water supplied. Table 9.1 lists the major objectionable ionic constituents present in many water supplies that can be removed by demineralization.

Prevention of scale and other deposits in cooling and boiling waters is best accomplished by removal of dissolved solids. Whereas in municipal water

TABLE 9.1   IONIC CONSTITUENTS IN WATER

Constituent	Chemical Designation	Problems Caused
Hardness	Calcium and magnesium salts expressed as $CaCO_2$, Ca, Mg.	Main source of scaling in heat-exchange equipment, boilers, pipe lines, and so on. Forms curds with soap, interferes with dyeing.
Alkalinity	Bicarbonate (HCO_3), carbonate (CO_3), and hydrate (OH), expressed as $CaCO_3$.	Cause of foaming and carryover of solids with steam. Embrittlement of boiler steel. Bicarbonate and carbonate produce CO_2 in steam, a source of corrosion.
Free Mineral Acidity	H_2SO_4, HCl, and so on, expressed as $CaCO_3$.	Rapid corrosion and deterioration.
Chloride	$Cl -$	Interferes with silvering processes and increases TDs.
Sulfate	$(SO_4) - -$	Calcium sulfate scale is formed.
Iron and Manganese	Fe + + (ferrous) Fe + + + (ferric) Mn + +	Discolors water, deposits in water lines, boilers, and so on. Interferes with dying, tanning, paper manufacture, and process work.
Carbon Dioxide	CO_2	Corrodes water lines, particularly steam and condensate lines.
Silica	SiO_2	Scale in boilers and cooling-water systems. Insoluble scale on turbine blades due to silica vaporization in high-pressure boilers (over 600 psi).

purification such removal is limited to the partial reduction of hardness and the removal of iron and manganese, in industrial water treatment it is often carried much further and may include the complete removal of hardness, the reduction or removal of alkalinity, the removal of silica, or even the complete removal of all dissolved solids.

The two most frequently encountered water problems—scale formation and corrosion—are common to cooling, heating, and steam-generating systems. Hardness (calcium and magnesium), alkalinity, sulfate, and silica all form the main source of scaling in heat-exchange equipment, boilers, and pipes. Scales or deposits formed in boilers and other exchange equipment act

as insulation, preventing efficient heat transfer and causing boiler tube failures through overheating of the metal. Free mineral acids (sulfates and
chlorides) cause rapid corrosion of boilers, heaters, and other metal containers
and piping. Alkalinity causes embrittlement of boiler steel, and carbon dioxide
and oxygen cause corrosion, primarily in steam and condensate lines.

Low-quality steam can produce undesirable deposits of salts and alkali
on the blades of steam turbines; much more difficult to remove are silica
deposits which can form on turbine blades even when steam is satisfactory
by ordinary standards. At steam pressures above 600 psi, silica from the boiler
water actually dissolves in the gaseous steam and then reprecipitates on the
turbine blades at their lower-pressure end.

In the operation of every cooling, heating, and steam-generating system,
the water changes temperature. Higher temperatures, of course, increase both
corrosion rates and scale-forming tendency. Evaporation in process steam
boilers and in evaporative cooling equipment increases the dissolved-solids
concentration of the water, compounding the problem.

In addition to the formation of scale or corrosion of metal within boilers,
auxiliary equipment is also susceptible to similar damage. Attempts to prevent
scale formation within a boiler can lead to makeup line deposits if the treatment chemicals are improperly chosen. Thus, the addition of normal phosphates to an unsoftened feed water can cause a dangerous condition by clog-
ging the makeup line with precipitated calcium phosphate. Deposits in the
form of calcium or magnesium stearate deposits, otherwise known as "bath-
tub ring" can be readily seen, and are caused by the combination of calcium
or magnesium with negative ions of soap stearates. Table 9.1 shows some
common ionic constituents in water.

THEORY

Ion exchangers are materials that can exchange one ion for another, hold it
temporarily, and release it to a regenerant solution. In a typical demineralizer,
this is accomplished in the following manner: The influent water is usually
passed through a hydrogen cation-exchange resin which converts the influent
salt (say, sodium sulfate) to the corresponding acid (say, sulfuric acid) by
exchanging an equivalent number of hydrogen ($H+$) ions for the metallic
cations ($Ca++$, $Mg++$, $Na+$). These acids are then removed by passing
the effluent through an alkali regenerated anion-exchange resin which replaces the anions in solution ($Cl-$, $SO_4=$, NO_3-) with an equivalent number
of hydroxide ions. The hydrogen ions and hydroxide ions neutralize each
other to form an equivalent amount of pure water (see Figure 9.1). During
regeneration, the reverse reaction takes place. The cation resin is regenerated
with either sulfuric or hydrochloric acid and the anion resin is regenerated
with sodium hydroxide.

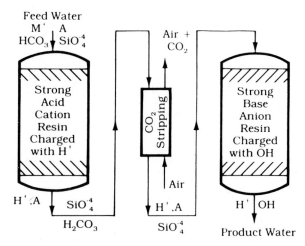

Figure 9.1 Ion exchange demineralization.

There are various arrangements or equipment possible but in all cases, except in mixed-bed demineralization, the water should first pass through a cation exchanger. In mixed demineralization, the two exchange materials (that is, the cation-exchange resin and the anion-exchange resin) are placed in one shell instead of two separate shells. In operation, the two types of exchange materials are thoroughly mixed so that we have, in effect, a number of multiple demineralizers in series. Higher-quality water is obtained from a mixed-bed unit than from a two-bed system. (see Figure 9.2). Operation of cation and anion exchanges is shown in Figure 9.3 (for fundamental processes) and Figure 9.4 (operation modes for both cation/anion exchanges).

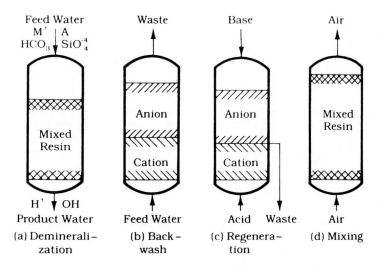

Figure 9.2 Mixed resin demineralization.

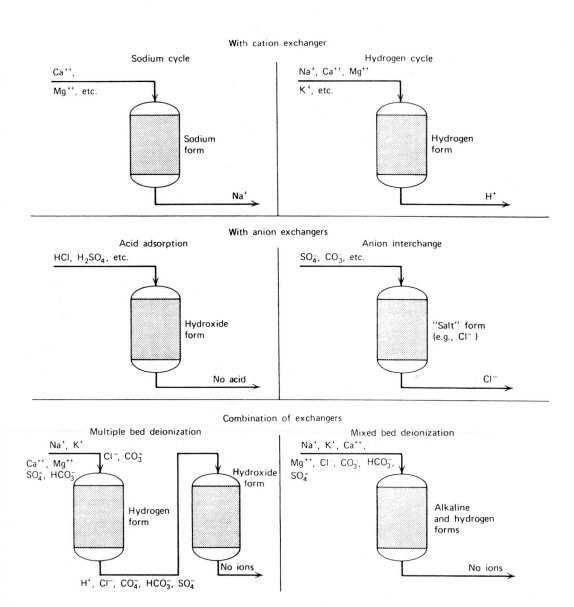

Figure 9.3 Fundamentals of ion exchange process.

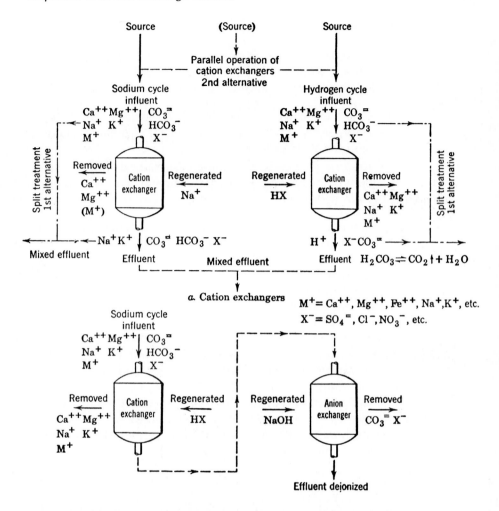

Figure 9.4 Operation of ion exchangers: (a) cation exchangers; (b) cation and anion exchangers.

PROPERTIES OF AN ION-EXCHANGE MATERIAL

To be suitable for industrial use, an ion-exchange resin must exhibit durable physical and chemical characteristics which are summarized by the following properties.

Functional Groups

The molecular structure of the resin is such that it must contain a macroreticular tissue with acid or basic radicals. These radicals are the basis of classifying ion exchangers into two general groups:

- Cation exchangers, in which the molecule contains acid radicals of the HSO_3 or HCO type able to fix mineral or organic cations and exchange with the hydrogen ion $H+$.
- Anion exchangers, containing basic radicals (for example, amine functions of the type NH_2) able to fix mineral or organic anions and exchange them with the hydroxyl ion $OH-$ coordinate to their dative bonds.

The presence of these radicals enable a cation exchanger to be assimilated to an acid of form H—R and an anion exchanger to be a base of form OH—R when regenerated.

These radicals act as immobile ion-exchange sites to which are attached the mobile cations or anions. For example, a typical sulfonic acid cation exchanger has immobile ion-exchange sites consisting of the anionic radicals SO to which are attached the mobile cations, such as $H+$ or $Na+$. An anion exchanger similarly has immobile cationic sites to which are attached mobile (exchangeable) hydroxide anions OH. The radicals attached to the molecular nucleus further determine the nature of the acid or base, whether it will be weak or strong.

Exchangers are divided into four specific classifications depending on the kind of radical, or functional group, attached; strong acid, strong base, weak acid, or weak base. Each of these four types of ion exchangers is described in detail later.

Solubility

The ion-exchange substance must be insoluble under normal conditions of use. All ion-exchange resins in current use are high molecular weight polyacids or polybases which are virtually insoluble in most aqueous and non-aqueous media. This is no longer true of some resins once a certain temperature has been reached. For example, some anion-exchange resins are limited to a maximum temperature of 105°F.

Liquid ion-exchange resins exist also, yet we do not consider their applicability here and they also exhibit very limited solubility in aqueous solutions.

Bead Size

The resins must be in the form of spherical granules of maximum homogeneity and dimensions so that they do not pack too much, the void volume among their interstices is constant for a given type, and the liquid head loss in percolation remains acceptable. Most ion-exchange resins occur as small beads or granules usually between 16 and 50 mesh in size.

Resistance to Fracture

The ion or ionized complexes that the resins are required to fix are of varied dimensions and weights. The swelling and contraction of the resin bead that this causes must obviously not cause the grains to burst.

Another important factor is the bead resistance to osmotic shock which will inevitably occur across its boundary surface, as there will be a salinity gradient of different magnitude during the cycle of the exchange material. The design of ion-exchange apparatus also must take into consideration the safe operation of the ion-exchange resin and avoid excessive stresses or mechanical abrasion in the bed, which could lead to breakage of the beads.

TYPES OF ION-EXCHANGE RESINS

As mentioned previously, ion-exchange materials are grouped into four specific classifications depending on the functional group attached; strong-acid cation, strong-base anion, weak-acid cation, or weak-base anion. (See Figures 9.5 through 9.10.) In addition to these, we also have inert resins that do not have chemical properties.

Strongly Acidic Cation

Strongly acidic cation resins derive their exchange activity from sulfonic functional groups (HSO). The major cations in water are calcium, magnesium, sodium, and potassium and they are exchanged for hydrogen in the strong-acid cation exchanger when operated in the hydrogen cycle. The following equation represents the exhaustion phase, is written in the molecular form (as if the salts present were undissociated), and shows the cations in combination with the major anions, the bicarbonate, sulfate, and chloride anions:

$$\begin{matrix} Ca & 2HCO_2 & & Ca & 2H_2CO_3 \\ Mg & \cdot\ SO_4\ +\ 2RSO_3H & \rightleftharpoons & 2RSO_3\ Mg\ + & H_2SO_4 \\ Na & 2Cl & & Na & 2HCl \end{matrix}$$

where R represents the complex resin matrix.

Because these equilibrium reactions are reversible, when the resin capacity has been exhausted it can be recovered through regeneration with a mineral acid. The strong-acid exchangers operate at any pH, split strong or weak salts, require excess strong-acid regenerant (typical regeneration efficiency varies from 25 percent to 45 percent in concurrent regeneration), and they permit low leakage. In addition, they have rapid exchange rates, are stable, exhibit swelling less than 7 percent going from $Na+$ to $H+$ form, and may last 20 years or more with little loss of capacity.

These resins have found a wide range of application, being used on the sodium cycle for softening, and on the hydrogen cycle for softening, de-alkalization, and demineralization.

Weakly Acidic Cation

Weakly acidic cation-exchange resins have carboxylic groups (COOH) as the exchange sites. When operated on the hydrogen cycle, the weakly acidic resins are capable of removing only those cations equivalent to the amount of alkalinity present in the water, and most efficiently the hardness (calcium and magnesium) associated with alkalinity, according to these reactions:

$$\left.\begin{array}{c} Ca \\ Mg \\ 2Na \end{array}\right\}(HCO_3) + RCOOH \rightleftharpoons 2RCOO \begin{array}{c} Ca \\ Mg \\ 2Na \end{array} + H_2CO_4$$

These reactions are also reversible and permit acid regeneration to return the exhausted resin to the hydrogen form.

The resin is highly efficient, for it is regenerated with 110 percent of the stoichiometric amount of acid as compared to 200 percent to 300 percent for strong-acid cation-exchange resins. It can be regenerated with the waste acid from a strong-acid cation exchanger and there is little waste problem during the regeneration cycle. In order to prevent calcium sulfate precipitation when regenerated with H_2SO_4, it is usually regenerated stepwise with initial H_2SO_4 at .5 percent. The resins are subject to reduced capacity from increasing flow rate (above 2 gpm/ft), low temperatures, and/or a hardness-alkalinity ratio especially below 1.0.

Weakly acidic resins are used primarily for softening and dealkalization, frequently in conjunction with a strongly acidic polishing resin. Systems which use both resins profit from the regeneration economy of the weakly acidic resin and produce treated water of quality comparable to that available with a strongly acidic resin.

Strongly Basic Anion

Strongly basic anion-exchange resins derive their functionality from the quaternary ammonium exchange sites. All the strongly basic resins used for demineralization purposes belong to two main groups commonly known as type I and type II. The principal difference between the two resins, opera-tionally, is that type I has a greater chemical stability, and type II has a slightly greater regeneration efficiency and capacity. Physically, the two types differ by the species of quaternary ammonium exchange sites they exhibit. Type I sites have three methyl groups, while in type II, an ethanol group replaces one of the methyl groups. In the hydroxide form, the strongly basic anion will remove all the commonly encountered inorganic acids according to three reactions:

$$\left.\begin{array}{l} H_2SO_4 \\ 2HCl \\ H_2SiO_3 \\ H_2CO_3 \end{array}\right\} + 2ZOH \rightleftharpoons \begin{array}{l} SO_4 \\ 2Cl \\ 2HSiO_3 \\ 2HCO_3 \end{array} + H_2O$$

Like the cation resin reactions, the anion-exchange reactions are also reversible and regeneration with a strong alkali, such as caustic soda, will return the resin to the hydroxide form.

The strong-base exchangers operate at any pH, can split strong or weak salts, require excess high-grade NaOH for regeneration (with the typical efficiency varying from 18 percent to 33 percent), are subject to organic fouling from such compounds when present in the raw water and to resin degradation due to oxidation and chemical breakdown. The strong-base anion resins suffer from capacity decrease and silica leakage increase at flow rates above 2 gpm/ft^3 of resin, and cannot operate over 130° to 150°F depending on resin type. The normal maximum continuous operating temperature is 120°F, and to minimize silica leakage, warm caustic (up to 120°F) should be used. Type I exchangers are for maximum silica removal. They are more difficult to regenerate and swell more (from Cl to OH form) than type II. The major case for selecting a type I resin is where high operating temperatures and/or very high silica levels are present in the influent water or superior resistance to oxidation or organics is required.

Type II exchangers remove silica (but less efficiently than type I) and other weak anions, regenerate more easily, are less subject to fouling, are freer from the odor of amine, and are cheaper to operate than type I. Where free mineral acids are the main constituent to be removed and very high silica removal is not required, type II anion resin should be chosen.

Weakly Basic Anion

Weakly basic anion resins derive their functionality from primary (R—NH), secondary (R—NHR'), tertiary (R—N—R'2), and sometimes quaternary amine groups. The weakly basic resin readily absorbs such free mineral acids as hydrochloric and sulfuric, and the reactions may be represented according to the following:

$$\begin{array}{l} H_2SO_4 + 2ZOH = 2Z\ SO_4 + 2HO \\ 2HCl \qquad\qquad\quad 2Cl \end{array}$$

Because the preceding reactions are also reversible, the weakly basic resins can be regenerated by applying caustic soda, soda ash, or ammonia. The weak-base exchanger regenerates with a nearly stoichiometric amount of base (with the regeneration efficiency possibly exceeding 90 percent) and can utilize waste caustic following strong-base anion-exchange resins. Weakly

TABLE 9.2 SODIUM CATION-EXCHANGER REACTIONS

Softening

$$Na_2Z + \frac{Ca}{Mg}\begin{Bmatrix}(HCO_3)_2\\SO_4\\Cl_2\end{Bmatrix} = \frac{Ca}{Mg}Z + \begin{Bmatrix}2NaHCO_3\\Na_2SO_4\\2NaCl\end{Bmatrix}$$

Sodium Cation Exchanger *(insoluble)* + Calcium and/or Magnesium {Bicarbonates Sulfates and/or Chlorides} *(soluble)* = Calcium and/or Magnesium Cation Exchanger *(insoluble)* + {Sodium Bicarbonate Sodium Sulfate and/or Sodium Chloride} *(soluble)*

Regeneration

$$\frac{Ca}{Mg}Z + 2NaCl = Na_2Z + \frac{Ca}{Mg}Cl_2$$

Calcium and/or Magnesium Cation Exchanger *(insoluble)* + Sodium Chloride *(soluble)* = Sodium Cation Exchanger *(insoluble)* + Calcium and/or Magnesium Chlorides *(soluble)*

The symbol Z represents sodium cation exchanger.

COMPENSATED HARDNESS: The hardness of a water for softening by the zeolite process should be compensated when:

1. The total hardness (T.H.) is over 400 ppm as $CaCO_3$, or
2. The sodium salts (Na) are over 100 ppm as $CaCO_3$

Calculated compensated hardness as follows:

$$\text{Compensated Hardness (ppm)} = \text{Total Hardness (ppm)} \times \frac{9{,}000}{9{,}000 - \text{Total Cations (ppm)}}$$

Express compensated hardness as:

1. Next higher tenth of a grain up to 5.0 grains per gallon.
2. Next higher half of a grain from 5.0 to 10.0 grains per gallon.
3. Next higher grain above 10.0 grains per gallon.

SALT CONSUMPTION: The salt consumption with the sodium cation-exchange water softener ranges between 0.275 and 0.533 lbs. of salt per 1,000 grains of hardness, expressed as calcium carbonate, removed. This range is due to two factors: (1) the composition of the water and (2) the operating exchange value at which the exchange resin is to be worked. The lower salt consumptions may be attained with waters that are not excessively hard nor high in sodium salts and where the exchange resin is not worked at its maximum capacity.

284

TABLE 9.3 HYDROGEN CATION-EXCHANGER RESINS

Reactions with Bicarbonates

$$\left. \begin{array}{c} Ca \\ Mg \\ Na_2 \end{array} \right\} (HCO_3)_2 \; + \; H_2Z \; = \; \left. \begin{array}{c} Ca \\ Mg \\ Na_2 \end{array} \right\} Z \; + \; 2H_2O \; + \; 2CO_2$$

Calcium, magnesium and/or sodium Bicarbonate *(soluble)* + Hydrogen Cation Exchanger *(insoluble)* = Cation Exchanger *(insoluble)* Calcium, magnesium and/or sodium + Water + Carbon Dioxide *(soluble gas)*

Reactions with Sulfates or Chlorides

$$\left. \begin{array}{c} Ca \\ Mg \\ Na_2 \end{array} \right\} \left\{ \begin{array}{c} SO_4 \\ Cl_2 \end{array} \right. \; + \; H_2Z \; = \; \left. \begin{array}{c} Ca \\ Mg \\ Na_2 \end{array} \right\} Z \; + \; \left\{ \begin{array}{c} H_2SO_4 \\ 2HCl \end{array} \right.$$

Calcium magnesium and/or sodium Sulfates and/or chlorides *(soluble)* + Hydrogen Cation Exchanger *(insoluble)* = Calcium, magnesium and/or sodium Cation Exchanger *(insoluble)* + Sulfuric and/or hydrochloric acids *(soluble)*

Regeneration Reactions

$$\left. \begin{array}{c} Ca \\ Mg \\ Na_2 \end{array} \right\} Z \; + \; H_2SO_4 \; = \; H_2Z \; + \; \left. \begin{array}{c} Ca \\ Mg \\ Na_2 \end{array} \right\} SO_4$$

Calcium, magnesium and/or sodium Cation Exchanger *(insoluble)* + Sulfuric Acid *(soluble)* = Hydrogen Cation Exchanger *(insoluble)* + Calcium, magnesium and/or sodium Sulfates *(soluble)*

The symbol Z represents hydrogen cation-exchanger radical.

basic resins are used for high strong-acid waters (Cl, SO_4, NO_3), and low alkalinity, do not remove anions satisfactorily above pH 6, do not remove CO or silica, but have capacities about twice as great as for strong-base exchangers. Weak-base resins can be used to precede a strong-base anion resin to provide the maximum protection of the latter against organic fouling and to reduce regenerant costs. Tables 9.2, 9.3, 9.4, and 9.5 indicate the exchange.

Inert Resin

There also exists a type of resin with no functional groups attached. This resin offers no capacity to the system but increases regeneration efficiency in mixed-bed exchangers. These inert resins are of a density between cation and anion resins and when present in mixed-bed vessels help to separate cation and anion resins during backwash.

TABLE 9.4   ANION-EXCHANGER REACTIONS (Weakly basic and strongly basic exchangers)

Reaction with Strongly Ionized Acids

$$R_3N \quad + \begin{cases} H_2SO_4 \\ 2HCL \end{cases} = \quad R_3N \cdot \begin{cases} H_2SO_4 \\ 2HCL \end{cases}$$

weakly basic Anion Exchanger $+$ $\begin{cases} \text{Sulfuric} \\ \text{and/or} \\ \text{hydrochloric acids} \end{cases}$ = Anion Exchanger $\begin{cases} \text{sulfate} \\ \text{and/or} \\ \text{hydrochloride} \end{cases}$

  (insoluble)   *(soluble)*   *(insoluble)*

Regeneration Reaction

$$R_3N \cdot H_2SO_4 \quad + \quad Na_2CO_3 = \quad R_3N \quad + Na_2SO_4 + \quad CO_2 \quad + H_2O$$

Anion Exchanger hydrosulfate $+$ soda ash $=$ weakly basic Anion Exchanger $+$ sodium sulfate $+$ carbon dioxide $+$ water

  (insoluble)   *(soluble)*   *(insoluble)*   *(soluble)*   *(soluble gas)*

Reaction with Weakly Ionized Acids

$$R_4NOH \quad + \quad H_2SiO_3 \quad = \quad R_4N \cdot HSiO_3 \quad + H_2O$$

strongly basic Anion Exchanger $+$ silicic acid $=$ Anion Exchanger silicate $+$ water

  (insoluble)   *(soluble)*   *(insoluble)*

Regeneration Reaction

$$(R_4N)_2SO_4 \quad + \quad 2NaOH \quad = \quad 2R_4NOH \quad + Na_2SO_4$$

Anion Exchanger sulfate $+$ caustic soda $=$ strongly basic Anion Exchanger $+$ sodium sulfate

  (insoluble)   *(soluble)*   *(insoluble)*   *(soluble)*

The symbol R_3N represents the complex weakly basic anion-exchanger radical.
The symbol R_4N represents the complex strongly basic anion-exchanger radical.

TABLE 9.5 REMOVAL AND RECOVERY—Ion-Exchange Reactions

Rinse Water Recovery

(a) $\quad H_2Cr_2O_7 + Al_2(Cr_2O_7)_3 + \quad 6RH \quad \rightarrow \quad 2R_3Al \quad + 4H_2Cr_2O_7$

$$\underset{\substack{\text{Chromic} \\ \text{Acid}}}{} + \underset{\substack{\text{Aluminum} \\ \text{Dichromate}}}{} + \underset{\substack{\text{Hydrogen} \\ \text{Cation} \\ \text{Exchanger}}}{} \quad \underset{\substack{\text{Aluminum} \\ \text{Cation} \\ \text{Exchanger}}}{} + \underset{\substack{\text{Chromic} \\ \text{Acid}}}{}$$

(b) $\quad H_2Cr_2O_7 + \quad 2R_4NOH \quad \rightarrow \quad (R_4N)_2Cr_2O_7 \quad + 2H_2O$

$$\underset{\substack{\text{Chromic} \\ \text{Acid}}}{} + \underset{\substack{\text{Hydroxide} \\ \text{Strongly Basic} \\ \text{Anion Exchanger}}}{} \quad \underset{\substack{\text{Dichromate} \\ \text{Strongly Basic} \\ \text{Anion Exchanger}}}{} + \text{Water}$$

Regeneration of Strongly Basic Anion Exchanger

(a) $\quad 4NaOH \quad + \quad (R_4N)_2Cr_2O_7 \quad \rightarrow \quad 2R_4NOH \quad + 2Na_2CrO_4 + H_2O$

$$\underset{\substack{\text{Sodium} \\ \text{Hydroxide}}}{} + \underset{\substack{\text{Chromate} \\ \text{Strongly Basic} \\ \text{Anion Exchanger}}}{} \quad \underset{\substack{\text{Hydroxide} \\ \text{Strongly Basic} \\ \text{Anion Exchanger}}}{} + \underset{\substack{\text{Sodium} \\ \text{Chromate}}}{} + \text{Water}$$

(b) $\quad 2Na_2CrO_4 + \quad 4RH \quad \rightarrow \quad 4RNa \quad + H_2Cr_2O_7 + H_2O$

$$\underset{\substack{\text{Sodium} \\ \text{Chromate}}}{} + \underset{\substack{\text{Hydrogen} \\ \text{Cation} \\ \text{Exchanger}}}{} \quad \underset{\substack{\text{Sodium} \\ \text{Cation} \\ \text{Exchanger}}}{} + \underset{\substack{\text{Chromic} \\ \text{Acid}}}{} + \text{Water}$$

Plating or Anodizing Bath Recovery

(a) $\quad H_2Cr_2O_7 + Al_2(Cr_2O_7)_3 + \quad 6RH \quad \rightarrow \quad 2R_3Al \quad + 4H_2Cr_2O_7$

$$\underset{\substack{\text{Chromic} \\ \text{Acid}}}{} + \underset{\substack{\text{Aluminum} \\ \text{Dichromate}}}{} + \underset{\substack{\text{Hydrogen} \\ \text{Cation} \\ \text{Exchanger}}}{} \quad \underset{\substack{\text{Aluminum} \\ \text{Cation} \\ \text{Exchanger}}}{} + \underset{\substack{\text{Chromic} \\ \text{Acid}}}{}$$

(b) $\quad 2R_3Al \quad + 3H_2SO_4 \rightarrow \quad 6RH \quad + \quad Al_2(SO_4)_3$

$$\underset{\substack{\text{Aluminum} \\ \text{Cation} \\ \text{Exchanger}}}{} + \underset{\substack{\text{Sulfuric} \\ \text{Acid}}}{} \quad \underset{\substack{\text{Hydrogen} \\ \text{Cation} \\ \text{Exchanger}}}{} + \underset{\substack{\text{Aluminum} \\ \text{Sulfate}}}{}$$

The symbol R represents the cation-exchanger radical.
The symbol R_4N represents the complex strongly basic anion-exchanger radical.

Advantages of inert resin:

- Classify cation and anion resins so that little or no mixing of cation or anion resin occurs before regeneration, and a buffering mid-bed collection zone exists.
- Improve regeneration efficiency, thereby reducing resin quantities needed.
- Protect against osmotic shock since the inert layer effectively prevents the exposure of cation resin to the caustic regenerant solution and the exposure of anion resin to the acid regenerant solution.

RESIN PERFORMANCE

Variances in resin performance and capacities can be expected from normal annual attrition rates of ion-exchange resins. Typical attrition losses that can be expected are given in the following list.

- Strong cation resin: 3 percent per year for three years or 1,000,000 gals/ cu. ft.
- Strong anion resin: 25 percent per year for two years or 1,000,000 gals/ cu. ft.
- Weak cation/anion: 10 percent per year for two years or 750,000 gals/ cu. ft.

A steady falloff of resin-exchange capacity is a matter of concern to the operator and is due to several conditions:

Improper backwash. Blowoff of resin from the vessel during the backwash step can occur if too high a backwash flow rate is used. This flow rate is temperature dependent and must be regulated accordingly. Also, adequate time must be allotted for backwashing to insure a clean bed prior to chemical injection.

Channeling. Cleavage and furrowing of the resin bed can be caused by faulty operational procedures or a clogged bed or underdrain. This can mean that the solution being treated follows the path of least resistance, runs through these furrows, and fails to contact active groups in other parts of the bed.

Incorrect chemical application. Resin capacities can suffer when the regenerant is applied in a concentration that is too high or too low. Another important parameter to be considered during chemical application is the location of the regenerant distributor. Excessive dilution of the regenerant chemical can occur in the vessel if the distributor is located too high above the resin bed. A recommended height is 3 inches above the bed level.

Mechanical strain. When broken beads and fines migrate to the top of the resin bed during service, mechanical strain is caused which results in channeling, increased pressure drop, or premature breakthrough. The combination of these resulting conditions leads to a drop in capacity.

Resin Fouling. In addition to the physical causes of capacity losses listed previously, there are a number of chemically caused problems that merit attention, specifically the several forms of resin fouling that may be found.

Organic fouling occurs on anion resins when organics precipitate onto basic exchange sites. Regeneration efficiency is then lowered, thereby reducing the exchange capacity of the resin. Causes of organic fouling are fulvic, humic, or tannic acids or degradation products of DVB (divinylbenzene) cross-linkage material of cation resins. The DVB is degraded through oxidation and causes irreversible fouling of downstream anion resins.

Iron fouling is caused by both forms of iron ions; the insoluble form will coat the resin bead surface and the soluble form can exchange and attach to exchange sites on the resin bead. These exchanged ions can be oxidized by subsequent cycles and precipitate ferric oxide within the bead interior.

Silica fouling is the accumulation of insoluble silica on anion resins. It is caused by improper regeneration which allows the silicate (ionic form) to hydrolyze to soluble silicic acid which in turn polymerizes to form colloidal silicic acid with the beads. Silica fouling occurs in weak-base anion resins when they are regenerated with silica-laden waste caustic from the strong-base anion resin unless intermediate partial dumping is done.

Microbiological fouling (MB) becomes a potential problem when microbic growth is supported by organic compounds, ammonia, nitrates, and so on which are concentrated on the resin. Signs of MB fouling are increased pressure drops, plugged distributor laterals, and highly contaminated treated water.

Calcium sulfate fouling occurs when sulfuric acid is used to regenerate a cation exchanger after exhaustion by a water high in calcium. The precipitate of calcium sulfate (gypsum) that forms can cause calcium and sulfate leakage during subsequent service runs. Given a sufficient calcium input in the water to treat, calcium sulfate fouling is especially prevalent when the percent solution of regenerant is greater than 5 percent, or the temperature is greater than 100°F, or when the flow rate is less than 1 gpm/cu ft. Stepwise injection of sulfuric acid during regeneration can help prevent fouling.

Aluminum fouling of resins can appear when aluminum floc from alum or other coagulants in pretreatment are encountered by the resin bead. This floc coats the resin bead and in the ionic form will be exchanged. However, these ions are not efficiently removed during regeneration so the available exchange sites continuously decrease in number.

Copper fouling is found primarily in condensate polishing applications. Capacity loss is due to copper oxides coating the resin beads.

Oil fouling does not cause chemical degradation but gives loss of capacity due to filming on the resin beads and the reduction of their active surface. Agglomeration of beads also occurs causing increased pressure drop, channeling, and premature breakthrough. The oil-fouling problem can be alleviated by the use of surfactants.

SEQUENCE OF OPERATION

The mode of operation for ion-exchange units can vary greatly from one system to the next, depending on the user's requirements. Service and regeneration cycles can be fully manual to totally automatic, with the method of regeneration being cocurrent, countercurrent, or external.

The most common means of ion exchange found in industry is the co-current downflow fixed-bed technique. Figures 9.11 and 9.12 illustrate the service run, backwash cycle, regenerant introduction cycle, and slow and fast rinse step of a cocurrent system. These are followed by Figure 9.13 depicting the service and regeneration cycles of a countercurrent unit.

The exhaustion phase is called the service run. This is followed by the regeneration phase which is necessary to bring the bed back to initial conditions to cycle. The regeneration phase includes four steps: backwashing to clean the bed, introduction of the excess regenerant, a slow rinse or displacement step to push the regenerant slowly through the bed, and finally a fast rinse to remove the excess regenerant from the resin and elute the unwanted ions to waste.

Service Cycle

The service cycle is normally terminated by one or a combination of the following criteria:

- High effluent conductivity.
- Total gallons throughput.
- High-pressure drop.
- High silica.
- High sodium.
- Variations in pH.
- Termination of the service cycle can be manually or automatically initiated.

Backwash Cycle

Normally, the first step in the regeneration sequence is designed to reverse flow from the service cycle using sufficient volume and flow rate to develop proper bed expansion for the purpose of removing suspended material (crud) trapped in the ion-exchange bed during the service cycle. The backwash waste water is collected by the raw water inlet distributor and diverted to waste via value sequencing. Backwash rate and internal design should avoid potential loss of whole bead resin during the backwash step. (Lower water temperature means more viscous force and more expansion.)

Regenerant Introduction

This introduction of regenerant chemicals can be cocurrent or countercurrent depending on effluent requirements, operating cost, and so on. Regenerant dosages (pounds per cubic foot), concentrations, flow rate, and contact time are determined for each application. The regenerant distribution and collection system must provide uniform contact throughout the bed and should avoid regenerant hideout. Additional effluent purity is obtained with countercurrent systems since the final resin contact in the service will be the most highly regenerated resin in the bed, creating a polishing effect.

Displacement/Slow Rinse Cycle

The final steps in the regeneration sequence are generally terminated on acceptable quality. Displacement, which precedes the rinse step, is generally an extension of the regenerant introduction step. The displacement step is designed to give final contact with the resin, removing the bulk of the spent regenerant from the resin bed.

Fast Rinse Cycle

The fast rinse step is essentially the service cycle except that the effluent is diverted to waste until quality is proven. This final rinse is always in the same direction as the service flow. Therefore, in countercurrent systems the displacement flow and rinse flows will be in opposite directions.

SEQUENCE OF OPERATION—MIXED BED UNITS

In mixed-bed units, both the cation and the anion resins are mixed together thoroughly in the same vessel by compressed air. The cation and the anion resins being next to each other constitute an infinite number of cation and anion exchangers. The effluent quality obtainable from a well-designed and operated mixed-bed exchanger will readily produce demineralized water of conductivity less than 0.5 mmho and silica less than 10 ppb.

Service Cycle

As far as the mode of operation is concerned, the service cycle of a mixed-bed unit is very similar to a conventional two-bed system, in that water flows into the top of the vessel, down through the bed, and the purified effluent comes out the bottom. It is in the regeneration and the preparation of it that the mixed-bed differs from the two-bed equipment. The resins must be separated, regenerated separately, and remixed for the next service cycle.

Backwash Cycle

Prior to regeneration, the cation and the anion resins are separated by backwashing at a flow rate of 3.0 to 3.5 gpm/ft. The separation occurs because of the difference in the density of the two types of resin. The cation resin, being heavier, settles on the bottom, while the anion resin, being lighter, settles on top of the cation resin. After backwashing, the bed is allowed to settle down for 5 to 10 minutes and two clearly distinct layers are formed. After separation, the two resins are independently regenerated.

Regenerant Introduction

The anion resin is regenerated with caustic flowing downward from the distributor placed just above the bed, while the cation resin is regenerated with either hydrochloric or sulfuric acid, usually flowing upward. The spent acid and caustic are collected in the interface collector, situated at the interface of the two resins. The regenerant injection can be carried out simultaneously as described or sequentially. In sequential regeneration, the cation-resin regeneration should precede the anion-resin regeneration to prevent the possibility of calcium carbonate and magnesium hydroxide precipitation, which may occur because of the anion-regeneration waste coming in contact with the exhausted cation resin. If this precipitation occurs, it can foul the resins at the interface. This becomes very critical when only the mixed-bed exchanger is installed to demineralize the incoming raw water.

In the case of sequential regeneration, during the caustic and acid injection period, a blocking flow of the demineralized water is provided in the opposite direction of the regenerant injection. This is required to prevent the caustic from entering the cation resin and acid from entering the anion resin. When regeneration is carried out simultaneously, acid and caustic injection flows act like blocking flows to each other and no additional blocking flow with water is needed. In a few sequential-type regeneration systems, acid is injected to flow downward through the central interface collector which now also acts as an acid distributor.

Rinsing and Air Mix Cycles

After completion of the acid and caustic injection, both the cation and anion resins are rinsed slowly to remove the majority of the regenerant, without attempting to eliminate it completely. After the use of 7–10 gallons of slow rinse volume per cubic foot of each type of resin, the unit is drained to lower the water to a few inches above the resin bed. The resins are now remixed with an upflow of air. After remixing, the unit is filled completely with water flowing slowly from the top, to prevent anion-resin separation in the upper layers. The mixed-bed exchanger is then rinsed at fast flow rates. The conductivity of the effluent water may be very high for a few minutes and will then drop suddenly to the value usually observed in the service cycle. This phenomenon is characteristic of mixed beds and is due to the absorption of the remaining acid or caustic in different parts of the bed, by one or the other resin. This, no doubt, results in the loss of resin capacity, but this loss is negligible as compared to the length of the service cycle and the savings in the overall time required for regeneration.

ION EXCHANGE SOFTENING (SODIUM ZEOLITE SOFTENING)

This is one of the ion-exchange processes used in water purification. In this process, sodium ions from the solid phase are exchanged with the hardness ions from the aqueous phase.

Consider a bed of ion-exchange resin having sodium as the exchangeable ion, with water containing calcium and magnesium hardness allowed to percolate through this bed. Let us denote the ion-exchange resinous material as RNa, where R stands for resin matrix and Na is its mobile exchange ion. The hard water will exchange Ca and Mg ions rapidly, so that water at the effluent will be almost completely softened. Calcium and magnesium salts will be converted into corresponding sodium salts.

The reaction will proceed toward the right-hand side to its completion until the bed gets completely exhausted or saturated with Ca and Mg ions. In order to reverse the equilibrium so that the reaction proceeds toward the left-hand side, the concentration of sodium ions has to be increased. This increase in sodium ions is accomplished by using a brine solution of sufficient strength so that the total sodium ions present in the brine are more than the total equivalent of Ca and Mg in the exhausted bed. This reverse reaction is carried out in order to bring the exhausted resin back to its sodium form. This process is known as *regeneration*. When the softener with the fresh resin in sodium form is put in service, the sodium ions in the surface layer of the bed are immediately exchanged with calcium and magnesium, thereby producing soft water with very little residual hardness in the effluent. As the process continues, the resin bed keeps exchanging its sodium ions with calcium and magnesium ions until the hardness concentration increases rapidly and the softening run is ended.

This softening process can be extended to a point where the hardness coming in and going out is the same. When this condition is reached, the bed is completely exhausted and does not have any further capacity to exchange ions. This capacity is called the *total breakthrough capacity*. In practice, the softening process is never extended to reach this stage as it is ended at some predetermined effluent hardness, much lower than the influent hardness. This capacity is called the *operating exchange capacity*. After the resin bed has reached this capacity, the resin bed is regenerated with a brine solution.

The regeneration of the resin bed is never complete. Some traces of calcium and magnesium remain in the bed and are present in the lower-bed level. In the service run, sodium ions exchanged from the top layers of the bed form a very dilute regenerant solution which passes through the resin bed to the lower portion of the bed. This solution tends to leach some of the hardness ions not removed by previous regeneration. These hardness ions appear in the effluent water as *leakage*. Hardness leakage is also dependent on the raw water characteristics. If the Na/Ca ratio and calcium hardness are very high in the raw water, leakage of the hardness ions will be higher.

SEQUENCE OF OPERATION—SOFTENER UNITS

Following are the basic steps involved in a regeneration of a water softener.

Backwashing

After exhaustion, the bed is backwashed to effect a 50 percent minimum bed expansion to release any trapped air from the air pockets, minimize the compactness of the bed, reclassify the resin particles, and purge the bed of any suspended insoluble material. Backwashing is normally carried out at 5–6 gpm/ft. However, the backwash flow rates are directly proportional to the temperature of water.

Brine Injection

After backwashing, a 5 percent to 10 percent brine solution is injected during a 30-minute period. The maximum exchange capacity of the resin is restored with 10 percent strength of brine solution. The brine is injected through a separate distributor placed slightly above the resin bed.

Displacement or Slow Rinse

After brine injection, the salt solution remaining inside the vessel is displaced slowly, at the same rate as the brine injection rate. The slow rinsing should be continued for at least 15 minutes and the slow rinse volume should not be less than 10 gallons/cu ft of the resin. The actual duration of the slow rinse should be based on the greater of these two parameters.

Fast Rinse

Rinsing is carried out to remove excessive brine from the resin. The rinsing operation is generally stopped when the effluent chloride concentration is less than 5–10 ppm in excess of the influent chloride concentration and the hardness is equal to or less than 1 ppm as CaCO.

DEMINERALIZER ARRANGEMENTS

Each arrangement will vary substantially in both operating and installed costs. Important factors for selection are:

1. Influent water analysis.
2. Flow rate.
3. Effluent quality.
4. Waste requirements.
5. Operating cost.

Some typical arrangements (Figures 9.5 through 9.13) are shown on the following pages.

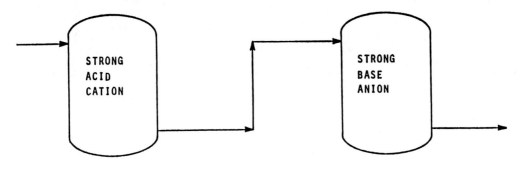

APPLICATION:

LOW ALKALINITY WATER WITH LOW EFFLUENT
SILICA LEVELS REQUIRED

-- HIGH OPERATING COSTS

-- YIELDS APPROXIMATELY NEUTRAL
   WASTE

Figure 9.5   Arrangement 1.

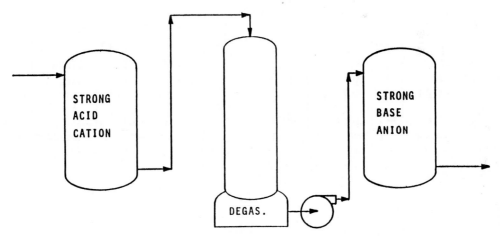

APPLICATION:

HIGH ALKALINITY WATER WITH LOW EFFLUENT
SILICA LEVELS REQUIRED.

-- HIGHER INITIAL COST BUT LOWER OPERATING
   COST THAN ARRANGEMENT 1

-- REQUIRES ADDITIONAL CHEMICALS TO
   NEUTRALIZE WASTE.

-- REPUMPING REQUIRED.

Figure 9.6   Arrangement 2.

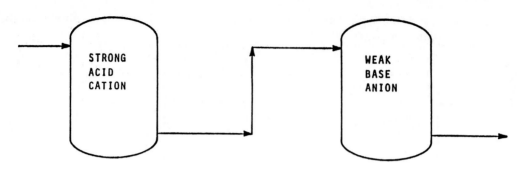

APPLICATION:

SAME AS ARRANGEMENT 1 EXCEPT NO SILICA
REMOVAL

-- EXCELLENT REMOVAL OF STRONG ACIDS
-- LOW OPERATING COST
-- REQUIRES ADDITIONAL CHEMICALS TO
   NEUTRALIZE WASTE

Figure 9.7   Arrangement 3.

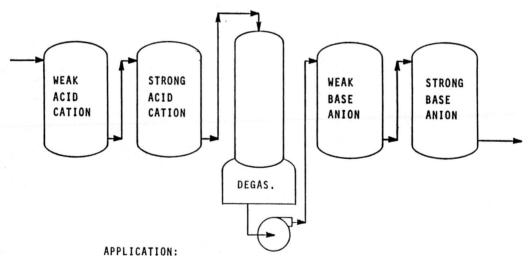

APPLICATION:

HIGH ALKALINITY, HIGH HARDNESS, HIGH CHLORIDE
AND SULFATE WATERS WITH HARDNESS TO ALKALINITY
RATIO   NEARING 1.0.

-- HIGH REGENERATION EFFICIENCES (CATIONS AND
   ANIONS) REGENERATED IN SERIES.

-- LOW SILICA EFFLUENT

-- REPUMPING REQUIRED

Figure 9.8   Arrangement 4.

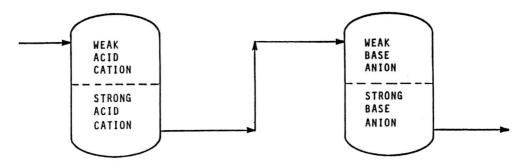

APPLICATION:

SAME AS ARRANGEMENT 4 BUT WITH LOWER ALKALINITY

-- LOW OPERATING AND LOW INITIAL COSTS
-- RESIN SEPARATION MUST BE MAINTAINED
-- DIFFICULT TO OPERATE
    -- REQUIRES PRECISE BACKWASH RATES TO MAINTAIN STATIFICATION
    -- SILICA POLYMERIZATION CAN OCCUR ON THE WEAK BASE ANION RESIN, LEADING TO EMBRITTLEMENT AND FRACTURE

Figure 9.9 Arrangement 5.

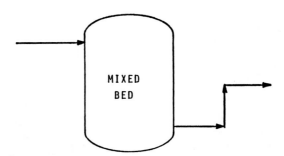

APPLICATION:

USED AS AN EFFLUENT POLISHER TO FURTHER REDUCE THE T.D.S. BEYOND THE LIMIT OF A PRECEDING SYSTEM.

-- YIELDS EXTRA PURE WATER

-- CATION AND ANION RESINS ARE ULTIMATELY MIXED DURING SERVICE, THEN STRATIFIED BY A BACKWASH PRIOR TO REGENERANT INTRODUCTION.

Figure 9.10 Arrangement 6.

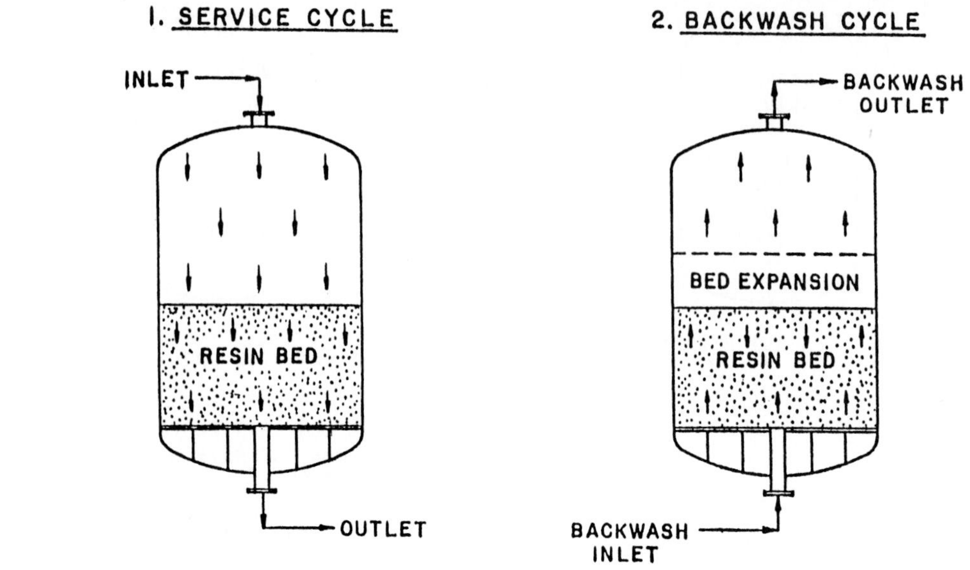

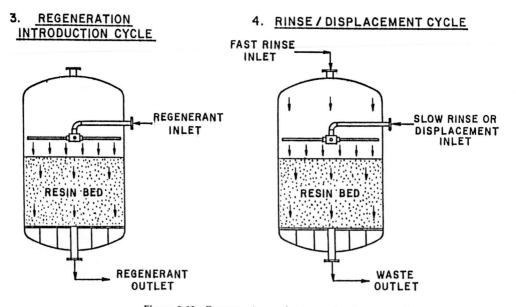

1. SERVICE CYCLE

INLET

RESIN BED

OUTLET

2. BACKWASH CYCLE

BACKWASH OUTLET

BED EXPANSION

RESIN BED

BACKWASH INLET

Figure 9.11   Service and backwash cycles diagrammed.

3. REGENERATION INTRODUCTION CYCLE

REGENERANT INLET

RESIN BED

REGENERANT OUTLET

4. RINSE / DISPLACEMENT CYCLE

FAST RINSE INLET

SLOW RINSE OR DISPLACEMENT INLET

RESIN BED

WASTE OUTLET

Figure 9.12   Regeneration and rinse cycles diagrammed.

SERVICE CYCLE (COUNTER CURRENT)

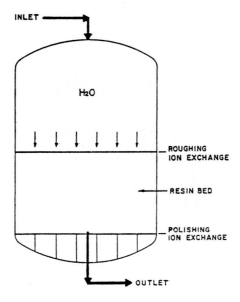

Figure 9.13 Service cycle diagrammed.

Terms Used in Ion-Exchange Technology

Acidity: An expression of the concentration of hydrogen ions present in a solution.

Adsorbent: A synthetic resin possessing the ability to attract and to hold charged particles.

Adsorption: The attachment of charged particles to the chemically active group on the surface and in the pores of an ion exchanger.

Alkalinity: An expression of the total basic anions (hydroxyl groups) present in a solution. It also represents, particularly in water analysis, the bicarbonate, carbonate, and occasionally, the borate, silicate, and phosphate salts which will react with water to produce the hydroxyl groups.

Anion: A negatively charged particle or ion.

Anion interchange: The displacement of one negatively charged particle by another on an anion-exchange material.

Attrition: The rubbing of one particle against another in a resin bed; frictional wear that will affect the site of resin particles.

Backwash: The countercurrent flow of water through a resin bed (that is, in at the bottom of the exchange unit, out at the top) to clean and regenerate the bed after exhaustion.

Base exchange: The property of the trading of cations shown by certain insoluble naturally occurring materials (zeolites) and developed to a high degree of specificity and efficiency in synthetic resin adsorbents.

Batch operation: The utilization of ion-exchange resins to treat a solution in a container wherein the removal of ions is accomplished by agitation of the solution and subsequent decanting of the treated liquid.

Bed: A mass of ion-exchange resin particles contained in a column.

Bed depth: The height of the resinous material in the column after the exchanger has been properly conditioned for effective operation.

Bed expansion: The effect produced during backwashing when the resin particles become separated and rise in the column. The expansion of the bed due to the increase in the space between resin particles may be controlled by regulating backwash flow.

Bicarbonate alkalinity: The presence in a solution of hydroxyl (OH—) ions resulting from the hydrolysis of carbonates or bicarbonates. When these salts react with water, a strong base and a weak acid are produced, and the solution is alkaline.

Breakthrough: The first appearance in the solution flowing from an ion-exchange unit of unabsorbed ions similar to those which are depleting the activity of the resin bed. Breakthrough is an indication that regeneration of the resin is necessary.

Capacity: The adsorption activity possessed in varying degrees by ion-exchange materials. This quality may be expressed as kilograins per cubic foot, gram-milliequivalents per gram, pound-equivalents per pound, gram-milliequivalents per milliliter, and so on, where the numerators of these ratios represent the weight of the ions adsorbed and the denominators represent the weight or volume of the adsorbent.

Carbonaceous exchangers: Ion-exchange materials of limited capacity prepared by the sulfonation of coal, lignite, peat, and so on.

Carboxylic: A term describing a specific acidic group (COOH) that contributes cation-exchange ability to some resins.

Cation: A positively charged particle or ion.

Channeling: Cleavage and furrowing of the bed due to faulty operational procedure, in which the solution being treated follows the path of least resistance, runs through these furrows, and fails to contact active groups in other parts of the bed.

Chemical stability: Resistance to chemical change which ion-exchange resins must possess despite contact with aggressive solutions.

Color-throw: Discoloration of the liquid passing through an ion-exchange material; the flushing from the resin interstices of traces of colored organic reaction intermediates.

Column operation: Conventional utilization of ion-exchange resins in columns through which pass, either upflow or downflow, the solution to be treated.

Cycle: A complete course of ion-exchange operation. For instance, a complete cycle of cation exchange would involve regeneration of the resin with acid, rinse to remove excess acid, exhaustion, backwash, and finally regeneration.

Deashing: The removal from solution of inorganic salts by means of adsorption by ion-exchange resins of both the cations and the anions that comprise the salts. See *deionization*.

Deionization: Deionization, a more general term than *deashing*, embraces the removal of all charged constituents or ionizable salts (both inorganic and organic) from solution.

Demineralizing: See *deashing*.

Density: The weight of a given volume of exchange material, backwashed and in place in the column.

Dissociation: Ionization.

Downflow: Conventional direction of solutions to be processed in ion-exchange column operation, that is, in at the top, out at the bottom of the column.

Dynamic system: An ion-exchange operation wherein a flow of the solution to be treated is involved.

Efficiency: The effectiveness of the operational performance of an ion exchanger. Efficiency in the adsorption of ions is expressed as the quantity of regenerant required to effect the removal of a specified unit weight of adsorbed material, for example, pounds of acid per kilogram of salt removed.

Effluent: The solution which emerges from an ion-exchange column.

Electrolyte: A chemical compound which dissociates or ionizes in water to produce a solution which will conduct an electric current; an acid, base, or salt.

Elution: The stripping of adsorbed ions from an ion-exchange material by the use of solutions containing other ions in concentrations higher than those of the ions to be stripped.

Equilibrium reactions: The interaction of ionizable compounds in which the products obtained tend to revert to the substance from which they were formed until a balance is reached in which both reactants and pacts are present in definite ratios.

Equivalent weight: The molecular weight of any element or radical expressed as grams, pounds, and so on divided by the valence.

Exchange velocity: The rate with which one ion is displaced from an exchanger in favor of another.

Exhaustion: The state in which the adsorbent is no longer capable of useful ion exchange; the depletion of the exchanger's supply of available ions. The exhaustion point is determined arbitrarily in terms of: (1) a value in parts per million of ions in the effluent solution; and (2) the reduction in quality of the effluent water determined by a conductivity bridge which measures the resistance of the water to the flow of an electric current.

Fines: Extremely small particles of ion-exchange materials.

Flow rate: The volume of solution which passes through a given quantity of resin within a given time. Flow rate is usually expressed in terms of feet per minute per cubic foot of resin or as milliliters per minute per milliliter of resin.

Freeboard: The space provided above the resin bed in an ion-exchange column to allow for expansion of the bed during backwashing.

Grain: A unit of weight; 0.0648 grams; 0.000143 pounds.

Grains per gallon: An expression of concentration of material in solution. One grain per gallon is equivalent to 17.1 parts per million.

Gram: A unit of weight; 15.432 grains; 0.0022 pounds.

Gram-milliquivalents: The equivalent weight in grams, divided by 1,000.

Greensands: Naturally occurring materials, composed primarily of complex silicates, which possess ion-exchange properties.

Hardness: The scale-forming and lather-inhibiting qualities which water, high in calcium and magnesium ions, possesses.

Hardness as calcium carbonate: The expression ascribed to the value obtained when the hardness-forming salts are calculated in terms of equivalent quantities of calcium carbonate; a convenient method of reducing all salts to a common basic for comparison.

Head loss: The reduction in liquid pressure associated with the passage of a solution through a bed of exchange material; a measure of the resistance of a resin bed to the flow of the liquid passing through it.

Hydraulic classification: The rearrangement of resin particles in an ion-exchange unit. As the backwash water flows up through the resin bed, the particles are placed in a mobile condition wherein the larger particles settle and the smaller particles rise to the top of the bed.

Hydrogen cycle: A complete course of cation-exchange operation in which the adsorbent is employed in the hydrogen or free acid form.

Hydroxyl: The term used to describe the anionic radical (OH—) which is responsible for the alkalinity of a solution.

Influent: The solution which enters an ion-exchange unit.

Ion: Any particle of less than colloidal size possessing either a positive or a negative electric charge.

Ionization: The dissociation of molecules into charged particles.

Ionization constant: An expression in absolute units of the extent of dissociation into ions of a chemical compound in solution.

Ion exchange: See fundamental description beginning page 273.

Kilograin: A unit of weight; 1,000 grains.

Leakage: The phenomenon in which some of the influent ions are not adsorbed and appear in the effluent when a solution is passed through an underregenerated exchange resin bed.

Negative charge: The electrical potential which an atom acquires when it gains one or more electrons; a characteristic of an anion.

pH: An expression of the acidity of a solution; the negative logarithm of the hydrogen-ion concentration (pH 1 very acidic; pH 14, very basic; pH 7, neutral).

pOH: An expression of the alkalinity of a solution; the negative logarithm of the hydroxyl-ion concentration.

pK: An expression of the extent of dissociation of an electrolyte; the negative logarithm of the ionization constant of a compound.

Physical stability: The quality which an ion-exchange resin must possess to resist changes that might be caused by attrition, high temperatures, and other physical conditions.

Positive charge: The electrical potential acquired by an atom which has lost one or more electrons; a characteristic of a cation.

Raw water: Untreated water from wells or from surface sources.

Regenerant: The solution used to restore the activity of an ion exchanger. Acids are employed to restore a cation exchanger to its hydrogen form; brine solutions may be used to convert the cation exchanger to the sodium form. The anion exchanger may be rejuvenated by treatment with an alkaline solution.

Regeneration: Restoration of the activity of an ion exchanger by replacing the ions adsorbed from the treated solution by ions that were adsorbed initially on the resin.

Rejuvenation: See *regeneration*.

Reverse deionization: The use of an anion-exchange unit and a cation-exchange unit—in that order—to remove all ions from solution.

Rinse: The operation which follows regeneration; a flushing out of excess regenerant solution.

Siliceous gel zeolite: A synthetic, inorganic exchanger produced by the aqueous reaction of alkali with aluminum salts.

Static system: The batch-wise employment of ion-exchange resins, wherein (since ion exchange is an equilibrium reaction) a definite endpoint is reached in which a finite quantity of all the ions involved is present. Opposed to a dynamic, column-type operation.

Sulfonic: A specific acidic group (SO_3H) on which depends the exchange activity of certain cation adsorbents.

Swelling: The expansion of an ion-exchange W which occurs when the reactive groups on the resin are converted from one form to another.

Throughput volume: The amount of solution passed through an exchange W before exhaustion of the resin is reached.

Upflow: The operation of an ion-exchange unit in which solutions are passed in at the bottom and out at the top of the container.

Voids: The space between the resinous particles in an ion-exchange bed.

Zeolite: Naturally occurring hydrous silicates exhibiting limited base exchange.

Index